Introduction to Biofuels Engineering

Introduction to Biofuels Engineering

Mason Jones

STATES
ACADEMIC PRESS
www.statesacademicpress.com

Introduction to Biofuels Engineering
Mason Jones
ISBN: 978-1-63989-302-7 (Hardback)

STATES
ACADEMIC PRESS

Published by States Academic Press,
109 South 5th Street,
Brooklyn, NY 11249, USA

Cataloging-in-Publication Data

Introduction to biofuels engineering / Mason Jones.
p. cm.
Includes bibliographical references and index.
ISBN 978-1-63989-302-7
1. Biomass energy. 2. Engineering. I. Jones, Mason.
TP339 .B56 2022
662.88--dc23

For more information regarding States Academic Press and its products, please visit the publisher's website www.statesacademicpress.com

Table of Contents

Preface

Any fuel that has been obtained by processing biomass such as plant, algae or animal waste that cannot be used as food or feed is termed as a biofuel. It is often considered as a renewable source of energy since the materials involved in its production can be replenished easily. Biofuels can be liquid, solid or gaseous. Bioethanol and biodiesel are liquid biofuels which are produced by the fermentation of carbohydrates found in sugarcane and transesterification of oils or fats. Biofuels engineering analyzes environmental and economic factors involved in the production of fossil fuel alternatives and direct old scientific principles into new technological solutions. This book attempts to understand the diverse facets that fall under the discipline of biofuels engineering and how such concepts have practical applications. Most of the topics introduced herein cover new techniques and the applications of biofuels engineering. The book is appropriate for those seeking detailed information in this area.

A detailed account of the significant topics covered in this book is provided below:

Chapter 1- The fuels which are obtained from biomass are called biofuels. They are broadly categorized into first generation, second generation, third generation and fourth generation biofuels. This chapter has been carefully written to provide an introduction to these categories of biofuels as well as life cycle assessment of biofuels.

Chapter 2- Biofuels are classified into various types. A few such types are biogas, bio alcohol fuels, aviation biofuel, syngas and biodiesel. Biogas is produced anaerobically by organic matter and bio alcohol fuels are generally produced using aliphatic alcohols such as methanol, ethanol, propanol and butanol. All these diverse types of biofuels have been carefully analyzed in this chapter.

Chapter 3- Biofuels can be derived from a large variety of sources. Some of the different types of biofuels on the basis of their sources are sugarcane biofuel, soybean biofuel, camelina biofuel, rapeseed biodiesel, cellulosic biofuel and animal fat biodiesel. This chapter discusses in detail these biofuels as well as their sources.

Chapter 4- The production of biofuels involves various chemical reactions such as esterification and transesterification. Chemically, petrodiesel and biodiesel are very similar; the only difference is the presence of two oxygen atoms in biofuels. The topics elaborated in this chapter will help in gaining a better perspective about the chemistry and production of biofuels as well as the methods used in the production of biofuels.

Chapter 5- There are various standards in place which govern the production, storage, usage and transportation of different biofuels. This chapter closely examines the various biofuel standards and policies in place around the world such as biofuel policy in Europe and biofuel policy in Brazil to provide an extensive understanding of the subject.

Chapter 6- There are numerous issues which are studied in relation to biofuels. Some of the prominent issues are the environmental impact of biodiesel, deforestation and indirect land use change impacts of biofuels. The topics elaborated in this chapter will help in gaining a better perspective about these issues connected to biofuels.

It gives me an immense pleasure to thank our entire team for their efforts. Finally in the end, I would like to thank my family and colleagues who have been a great source of inspiration and support.

Mason Jones

1

Biofuels: An Introduction

The fuels which are obtained from biomass are called biofuels. They are broadly categorized into first generation, second generation, third generation and fourth generation biofuels. This chapter has been carefully written to provide an introduction to these categories of biofuels as well as life cycle assessment of biofuels.

Any hydrocarbon fuel that is produced from an organic matter (living or once living material) in a short period of time (days, weeks, or even months) is considered a biofuel. Biofuels may be solid, liquid or gaseous in nature.

- Solid: Wood, dried plant material, and manure.
- Liquid: Bioethanol and Biodiesel.
- Gaseous: Biogas.

These can be used to replace or can be used in addition to diesel, petrol or other fossil fuels for transport, stationary, portable and other applications. Also, they can be used to generate heat and electricity. Some of the main reasons for shifting to biofuels are the rising prices of oil, emission of the greenhouse gases from fossil fuels and the interest for obtaining fuel from agricultural crops for the benefit of farmers.

Categories of Biofuels

First generation biofuels:

- These are made from food sources such as sugar, starch, vegetable oil, or animal fats using conventional technology.
- Common first-generation biofuels include Bioalcohols, Biodiesel, Vegetable oil, Bioethers, Biogas.
- Though the process of conversion is easy, but use of food sources in the production of biofuels creates an imbalance in food economy, leading to increased food prices and hunger.

Second generation biofuels:

- These are produced from non-food crops or portions of food crops that are not edible and considered as wastes, e.g. stems, husks, wood chips, and fruit skins and peeling.

- Thermochemical reactions or biochemical conversion process is used for producing such fuels. Examples include cellulose ethanol, biodiesel.
- Though these fuels do not affect food economy, their production is quite complicated.
- Also, it is reported that these biofuels emit less greenhouse gases when compared to first generation biofuels.

Third generation biofuels:

- These are produced from micro-organisms like algae. Example-Butanol.
- Micro-organisms like algae can be grown using land and water unsuitable for food production, therefore reducing the strain on already depleted water sources.
- One disadvantage is that fertilizers used in the production of such crops lead to environment pollution.

Fourth generation biofuels:

- In the production of these fuels, crops that are genetically engineered to take in high amounts of carbon are grown and harvested as biomass.
- The crops are then converted into fuel using second generation techniques.
- The fuel is pre-combusted and the carbon is captured. Then the carbon is geo-sequestered, meaning that the carbon is stored in depleted oil or gas fields or in unmineable coal seams.
- Some of these fuels are considered as carbon negative as their production pulls out carbon from environment.

Major Types of Biofuels	
Bioethanol	• It is derived from corn and sugarcane using fermentation process. • A litre of ethanol contains approximately two thirds of the energy provided by a litre of petrol. • When mixed with petrol, it improves the combustion performance and lowers the emissions of carbon monoxide and sulphur oxide.
Biodiesel	• It is derived from vegetable oils like soybean oil or palm oil, vegetable waste oils, and animal fats by a biochemical process called "Transesterification." • It produces very less or no amount of harmful gases as compared to diesel. • It can be used as an alternative for the conventional diesel fuel.

Biogas	• It is produced by anaerobic decomposition of organic matter like sewage from animals and humans. • Major proportion of biogas is methane and carbon dioxide, though it also has small proportions of hydrogen sulfide, hydrogen, carbon monoxide and siloxanes. • It is commonly used for heating, electricity and for automobiles.
Biobutanol	• It is produced in the same way as bioethanol i.e.through the fermentation of starch. • The energy content in butanol is the highest among the other gasoline alternatives. It can be added to diesel to reduce emissions. • It serves as a solvent in textile industry and is also used as a base in perfumes.
Biohydrogen	• Biohydrogen, like biogas, can be produced using a number of processes such as pyrolysis, gasification or biological fermentation. • It can be the perfect alternative for fossil fuel.

Advantages of Biofuels

- Availability: Biofuels are produced from biomass and thus are renewable.
- Source material: Whereas oil is a limited resource that comes from specific materials, biofuels can be manufactured from a wide range of materials including crop waste, manure, and other byproducts.
- Environment Pollution: Biofuels do not release as much carbon as fossil fuels do but fertilizers that are used in the growing bio fuels lead to greenhouse emissions. Also, biofuels can help in managing the municipal solid wastes i.e. the waste can be converted into fuel.
- Security: Biofuels can be produced locally, which decreases the nation's dependence upon foreign energy. By reducing dependence on foreign fuel sources, countries can protect the integrity of their energy resources and make them safe from outside influences.
- Economic stimulation: Because biofuels are produced locally, biofuel manufacturing plants can employ hundreds or thousands of workers, creating new jobs in rural areas. Biofuel production will also increase the demand for suitable biofuel crops, providing economic stimulation to the agriculture industry.

Disadvantages of Biofuels

- Efficiency: Fossil Fuels produce more energy than some of the biofuels. E.g. 1 gallon of ethanol produces less energy as compared to 1 gallon of gasoline (a fossil fuel).
- Cost: Pumping fossil fuels from the ground is a difficult and expensive process leading to high costs. Production of biofuels require land, this impacts cost of biofuels as well as that of food crops. Also, though growing engineered biofuel

crops can benefit farmers commercially but the excess number of such crops can also lead to loss of biodiversity.

- Food shortages: There is concern that using valuable cropland to grow fuel crops could have an impact on the cost of food and could possibly lead to food shortages.
- Water use: Massive quantities of water are required for proper irrigation of biofuel crops as well as to manufacture the fuel, which could strain local and regional water resources.

International Initiatives on Sustainable Biofuels

Production of sustainable biofuels is required to reduce their impact on environment and economy.

- Roundtable on Sustainable Biomaterials (RSB): It is an international initiative which brings together farmers, companies, governments, non-governmental organizations, and scientists who are interested in the sustainability of biofuel production and distribution.

  In April 2011, it launched a set of comprehensive sustainability criteria - the "RSB Certification System." Biofuels producers that meet these criteria are able to show buyers and regulators that their product has been obtained without harming the environment or violating human rights.
- Sustainable Biofuels Consensus: It is an international initiative which calls upon governments, the private sector, and other stakeholders to take decisive action to ensure the sustainable trade, production, and use of biofuels.
- Bonsucro: It is an international not for-profit, multi-stakeholder organization established in 2008 to promote sustainable sugar cane. Its stated aim is to reduce 'the environmental and social impacts of sugarcane production while recognising the need for economic viability'. It does that through setting sustainability standards and certifying sugar cane products including ethanol, sugar and molasses.

Bioenergy

Organic matter used to produce bioenergy is called biomass or bioenergy feedstocks. Bioenergy can be traced back to energy from sunlight, produced via photosynthesis, making it a major renewable energy source. As a storage house of bioenergy, biomass can be considered to be natures 'solar batteries'.

The energy biomass produces can be converted into electricity, heat or biofuels. Bioenergy can be as simple as a log fire or as complex as an advanced second generation liquid biofuel. Bioenergy is the most widely used renewable energy in the world, providing around 10% of the world's primary energy supplies, mostly as thermal energy for heating and cooking.

Types of Biomass

Sources of biomass include agricultural crops, animal and plant wastes, algae, wood and organic residential/industrial waste. The type of biomass will determine the type and amount of bioenergy that can be produced and the technology that can be used to produce it.

For example, agricultural crops, like corn and canola, can be used to produce liquid biofuels such as ethanol and biodiesel. Alternatively, wet wastes like manure are well suited to produce biogas through anaerobic digestion, which can be combusted to generate electricity and heat or upgraded into a transport fuel, biomethane.

How is Bioenergy Produced?

There are many ways to produce bioenergy. Choosing the best pathway and technology depends on the biomass material and the type of bioenergy you want to produce. Some processes can be relatively simple, like growing, harvesting and burning wood for heat generation. Other complex processes, like algae production for transport fuels, require a controlled growing environment using specific microalgae species. The algae are then processed to separate the oils which are refined into biofuels.

Fig: Feeding briquettes into biomass boiler.

A variety of conversion pathways can be used to convert biomass into bioenergy in the form of heat, electricity, or transportation fuels. Biomass conversion pathways include thermal, biochemical or mechanical either alone or in combination. Biomass can be converted to energy via a range of technologies from simple solid wood combustion heaters, to boilers and biodigesters which in turn produce gas for process steam for heat or for powering engines and turbines.

Biorefineries are facilities used to convert biomass into a number of fuel types and other bio-products; similar to conventional oil refineries. Fuel cells can convert hydrogen, produced from biomass, into electricity for stationary or transport energy.

Biomass Conversion Technologies

Biomass Conversion Technologies for Stationary Electricity and Heat Generation

There is a wide range of feedstocks that can be used, and technologies and processes for extracting energy from biomass and converting it into stationary bioenergy for heat and/or electricity. The more commonly used being conventional combustion, gasification, pyrolysis and anaerobic digestion.

Conventional Combustion

Direct combustion is the simplest and most widely used bioenergy technology for converting biomass to heat which can then be used for space heating or cooling, to heat water, for use in industrial processes, or to produce electricity via a steam engine or turbine. Combustion typically has an electrical efficiency of only 20-35%, but co-generation techniques can reach energy efficiencies of over 85%.

Fig: Woodlands biomass power plant.

Modern thermal-only bioenergy systems typically operate at or above 85% efficiency and are much simpler and cheaper to install and run than co-generation systems. Worldwide, thermal systems are far more common than biomass co-generation systems. The two main combustion technologies are:

- Fixed bed combustion: A common technology which involves burning materials on a fixed or moving grate with air passing through it.

- Fluidised bed combustion: Biomass is mixed with sand which then acts more like a fluid, burning more evenly and leading to increased efficiencies. It also allows a wider range of fuel types and higher moisture contents. Generally, fluidised bed boilers produce lower emissions than fixed bed boilers.

Co-firing is where biomass fuels, such as sawdust, biomass pellets, or biogas are combined and burnt with another base fuel, such as coal or LPG. Co-firing can be a cost-effective way for fossil fuel power generators to reduce greenhouse gas (GHG) emissions. Various fuels can be combusted with coal with minimal processing beyond chipping or shredding and drying. Most fuels are easier to handle, transport and store if they are processed and compacted into pellets or briquettes, which are denser and more consistent in quality and moisture content.

Co-generation, also known as Combined Heat and Power (CHP) has greater energy conversion efficiencies because it captures 'waste' heat from electricity generation, which can be used for space and water heating, or cooling via absorption chillers. Co-generation is well suited in situations where heating or cooling requirements are constant and where electricity can be used on site. Co-generation has traditionally been used to capture waste heat from conventional steam turbines.

Combined cycle electricity generation describes gas turbine systems that capture exhaust gases to heat water and produce steam which can then be used to drive a steam turbine to generate more power. The principle is that the exhaust of one gas engine is used as the heat source for another, thus extracting more useful energy from the heat, increasing the system's overall efficiency.

Tri-generation technology adds cooling to the co- generation process. Waste heat can be used in cooling via:

- Absorption chilling/refrigeration: Where heat drives a cooling system using a closed cycle of evaporating, dissolving and separating out two liquids at different pressures.
- Desiccant cooling: Using waste heat in a closed system to dry chemicals that are, in turn, used to extract moisture from the air before it is cooled in an evaporative cooler.

Gasification

Gasification is a thermo-chemical process that involves heating solid biomass to temperatures of around 800-1000 °C in a gasifier with a limited supply of oxygen. Under these conditions, fuel is only partly burnt and is largely converted to 'syngas' which contains a mixture of methane, hydrogen, carbon monoxide, carbon dioxide and nitrogen. Small amounts of char are produced through gasification.

Syngas can be used directly for heat or power applications, for example to run gas engines, gas turbines or combined cycle power systems. It can also be upgraded for biofuel production via a number of existing and emerging technologies.

Gasification is generally more efficient than combustion-based routes in terms of electricity generation. However, it is more demanding in terms of biomass specifications like moisture content and particle size. The need to scrub gases and dispose of tars can be an issue if the syngas is to be run through a gas engine to generate power.

Plasma gasification technology uses electricity in combination with an arc gasifier to create high temperatures which breaks down biomass and inorganic matter into syngas that can be used to generate electrical energy. At present, Plasma gasification is mainly deployed as a waste treatment technology focusing on inorganic waste streams including household rubbish.

Pyrolysis

Fig: Un-cleaned mustard seed.

Pyrolysis is similar to gasification, in that it involves thermal degradation of biomass heated in the absence of air - or with very limited air or oxygen. It produces solid, liquid and/or gaseous products at ratios dependent on the speed and temperature of the pyrolysis process. The gases and compounds in the liquids can be used to generate bioenergy.

- Slow pyrolysis involves heating biomass to temperatures of around 500°C and results in roughly equal proportions of biochar, liquid, (bio-oil) and syngas.
- Fast 'flash' pyrolysis is done at much higher temperatures, and can yield up to 80% bio-oil which can then be used in other stationary energy and biofuel production systems. Fast pyrolysis produces small quantities of syngas and biochar.
- Biochar is a stable form of charcoal that can be produced in many different ways

and thus have many varied chemical and structural properties. It may be used later to produce heat or it may have other commercial value as a soil improvement and carbon sequestration product.

Anaerobic Digestion

Anaerobic digestion is the biological breakdown of biomass in oxygen-free conditions. Anaerobic digestion occurs naturally e.g. in oxygen starved peat swamps and in man-made environments, including landfills, effluent lagoons and purpose-built biodigesters.

Fig: Dairy cows grazing pasture.

Many types of biomass can be anaerobically digested in biodigesters, but the process is particularly suited to wet feedstocks that do not contain lignin. Sewage, manure, wet agricultural residues, straw and effluents can be anaerobically digested to produce biogas - a mixture of mostly methane and carbon dioxide. Biogas can then be combusted to produce heat and/or power in a gas turbine. It can also be upgraded to natural gas standards and exported to the gas grid, or used in gas engine vehicles.

Fig: Biogas engines.

Biodigesters are sealed systems that intentionally promote the controlled build-up of biogas. They can be very simple in design and are commonly built and used by households in developing countries. There are also many larger systems used by farmers and food processors in countries such as Germany. The undigested sludge residue from biodigesters can be dehydrated and burnt to produce more bioenergy, or used as an organic fertiliser or compost.

Biogases can be captured in landfills and sewage treatment ponds through collection pipes and flared (burnt off) or used to generate bioenergy.

Advantages of Bioenergy

The carbon dioxide (CO_2) released from biomass during production of bioenergy is from carbon that circulates the atmosphere in a loop through the process of photosynthesis and decomposition. Therefore, production of bioenergy does not contribute extra CO_2 to the atmosphere like fossil fuels. Fossil fuels are a finite resource, developed through geological processes over millions of years and their use represents a one-way flow of GHGs from beneath the earth's surface to the atmosphere.

Reducing Greenhouse Gas Emissions by using Bioenergy

The extent of GHG emissions reduction varies widely and depends on many factors including the biomass (feedstocks) used how they are produced and procured and the type and efficiency of the technology used to produce bioenergy. Generally, GHG emissions reduction from bioenergy systems is greatest where waste biomass is converted to heat or combined heat and power in modern plants located near to where the waste is generated.

Bioenergy's GHG reduction benefits are potentially greater than those of other renewables. For example, stubble that is destined to be burnt in the field can be harvested and combusted in an emissions controlled bioenergy plant. Hence, GHG emissions reductions are made twice – once in the field through reduced burning and again through fossil fuel substitution from bioenergy production. Considerable research is underway around the world to quantify the total lifecycle impacts of various bioenergy and other renewable energy systems.

Renewable Energy

Bioenergy is a renewable energy that can generate many additional benefits, the extent of which depends on a combination of factors including the types of feedstocks used, how they are produced and transported and the efficiency of the technologies deployed to convert them to bioenergy.

Generating Heat and Electricity

Unlike most other renewable energy sources, biomass can generate both heat and electricity in a combined heat power (CHP) plant. This can then be used for a range of heating and cooling applications in industry, or for small communities.

Fig: Berrybank farm, Victoria which converts piggery waste to energy.

Bioenergy Qualities

- Better air quality: Bioenergy can provide air quality benefits where biomass residues that would otherwise be open-burnt in the field or forest, such as stubble, tree prunings or forest slash is removed and burnt in advanced emissions controlled bioenergy plant.
- Biofuels are biodegradable: Petroleum-based fuels and petrochemicals can be harmful to the environment and are major surface and ground water-pollutants. Biofuels such as ethanol and biodiesel are less toxic and are biodegradable.
- Regional and rural economic development and employment opportunities: International and Australian studies indicate that bioenergy creates many ongoing jobs; generally more than most other types of renewable energy. Bioenergy helps stimulate regional economic development and employment by providing new, decentralised and diversified income streams from bioenergy and biomass production. This gives landholders more market options for their traditional agricultural and tree crops and for their use of waste streams such as manures. It may also open up opportunities to grow new crops, especially on marginal or low rainfall farmland, e.g. Juncea for biodiesel as a low rainfall break-crop. New employment opportunities arise from growing and harvesting biomass, transport, handling, and through procurement, construction, operation and maintenance of bioenergy plants.
- Supporting agricultural and food-processing industries: Using biomass can help build resilience in agricultural, timber and food-processing industries. Bioenergy provides a use for their waste streams, can help them reduce their energy costs and potentially add a new revenue stream if they can sell biomass-derived heat and/or export 'green' electricity to the grid.

Using biomass can help build resilience in agricultural, timber and food-processing industries. Bioenergy provides a use for their waste streams, can help them reduce their energy costs and potentially add a new revenue stream if they can sell biomass-derived heat and/or export 'green' electricity to the grid.

- Cost savings: Using the right bioenergy technology in the right situation can help achieve greater cost savings than using fossil fuels. For example, areas that are reliant on LPG for heating (not linked to natural gas), areas remote from, or near the end of the power grid, subject to 'blackouts' and 'brownouts' and where electricity transmission losses and costs to upgrade the power supply are high.
- Less landfill: Using waste streams to generate bioenergy saves the environmental and economic costs of disposal in landfills and reduces contamination risks.
- Energy reliability and security: Rural and regional energy reliability and security can be enhanced by providing a domestic energy source that can run continuously, or at peak times as required by the electricity market, with greater flexibility to ramp up production at short notice than large coal-fired plants.
- A growing range of technologies and applications: There's a growing range of proven, adaptable technologies available for converting biomass into heat, electricity and biofuels. Bioenergy and biofuel production can link with the development of other bioproducts and biotechnologies. For example, organic digestates produced as a bi- product of anaerobic digestion, can be used as a fertiliser or soil enhancer. Biomass can produce useful chemicals as part of an integrated biorefinery system – similar to an oil refinery.
- Alternatives to prescribed forest burning: Bioenergy production can provide an alternative to prescribed burning of forests. Mechanical thinning and biomass removal for bioenergy can be used as a technique to reduce hazardous fuel levels, especially in areas where the cost and risks associated with prescribed burning are high.

- Water quality benefits have also been recorded where fuel reduction burning is replaced with biomass harvesting. For these reasons, biomass harvesting and removal for bioenergy and other small wood applications is a technique widely used in forests and woodlands in the USA.

- Environmental benefits from growing certain bioenergy crops: Bioenergy crops can be grown in areas that benefit from the additional vegetation cover. For example, trees can be grown and harvested for their woody biomass on farms in configurations that provide farm shelter, shade, salinity control, biodiversity and carbon sinks.

Fig: Plantation Deans Marsh.

Species such as Mallee eucalypts are widely grown in Australia and, due to their ability to coppice, (re-shoot), are able to be repeatedly harvested and regrown to provide renewable energy and other regional and on-farm benefits.

Logistics of Bioenergy

Because biomass is mostly available in low-density form, it demands more storage space, transport and handling than fossil equivalents, with consequent cost implications. Biomass often needs to be processed (pretreated) to improve handling. For most bioenergy systems and chains, handling and transport of biomass from the source location to the conversion plant is an important contributor to the overall costs of energy production. Crop harvesting, storage, transport, pretreatment and delivery can amount to 20 to 50% of the total costs of energy production.

Use of a single agricultural biomass feedstock for year-round energy generation requires relatively large storage because biomass is only available for a short time following harvest in many places. In addition to such seasonal variations in biomass availability, other characteristics complicate the biomass supply chain and should be taken into account. These include multiple feedstocks with their own complex supply chains, and storage challenges such as space constraints, fi re hazards, moisture control and health risks from fungi and spores.

Solid Biomass Supplies and Market Development for Utilization

Over time, several stages may be observed in biomass utilization and market developments in biomass supplies. Different countries seem to follow these stages over time, but clearly differ in their respective stages of development.

- Waste treatment (e.g., MSW and use of process residues (paper industry, food industry) onsite at production facilities) is generally the starting phase of a developing bioenergy system. Resources are available and often have a disposal cost (could have a negative value) making utilization profitable and simultaneously solving waste management problems. Large- and small-scale developments are evolving along with integrated resource management.

- Local utilization of resources from forest management and agriculture. Such resources are more expensive to collect and transport, but usually still economically attractive. Infrastructure development is needed.

- Biomass market development at regional scale; larger-scale conversion units with increasing fuel flexibility are deployed; increasing average transport distances further improves economies of scale. Increasing costs of biomass supplies make more energy-efficient conversion facilities necessary as well as feasible. Policy support measures such as feed-in tariffs (FITs) are usually needed to develop into this stage.

- Development of national markets with increasing numbers of suppliers and buyers; creation of a marketplace; increasingly complex logistics. Availability often increases due to improved supply systems and access to markets. Price levels may therefore decrease.

- Increasing scale of markets and transport distances, including cross-border transport of biofuels; international trade in biomass resources (and energy carriers derived from biomass). Biomass is increasingly becoming a globally traded energy commodity. Bio-ethanol trade has come closest to that situation.

- Growing role for dedicated fuel supply systems (biomass production largely or only for energy purposes). So far, most energy crops are grown because of agricultural interests and support (subsidies for farmers, use of set-aside subsidies), which concentrate on oil crops (such as rapeseed) and surplus food crops (cereals and sugar beets).

Countries that have gained substantial commercial experience with biomass supplies and biomass markets are generally able to obtain substantial cost reductions in biomass supply chains over time. In Finland and Sweden, delivery costs decreased from USD 2005 12 to 5/GJ from 1975 to 2003, due to factors such as scale increases, technological innovations or increased competition. Similar trends are observed in the corn ethanol industry in the USA and the sugarcane ethanol industry in Brazil.

Analyses of regional and international biomass supply chains show that road transport of untreated and bulky biomass becomes uncompetitive and energy-inefficient when crossing distances of 50 to 150 km. When long-distance transport is required, early pretreatment and densification in the supply chain pays off to minimize transport costs. Taking into account energy use and related GHG emissions, well-organized logistic chains can require less than 10% of the initial energy content of the biomass, but this requires substantial scale in transport, efficient pretreatment and minimization of road transport of untreated biomass.

Such organization is observed in the rapidly developing international wood pellet markets. Furthermore, (long distance) transport costs of liquid fuels such as ethanol and vegetable oils contribute only a minor fraction of overall costs and energy use of bioenergy chains.

Solid Biomass and Charcoal Supplies in Developing Countries

The majority of poorest households in the developing world depend on solid biomass fuels such as charcoal for cooking, and millions of small industries (such as brick and pottery kilns) generate process heat from these fuels. Despite this pivotal role of biomass, the sector remains largely unregulated, poorly understood, and the supply chains are predominantly in the hands of the informal sector.

When fuelwood is marketed, trees are usually felled and cut into large pieces and transported to local storage facilities where they are collected by merchants and delivered to wholesale and retail facilities, mainly in rural areas. Some of the wood is converted to charcoal in kilns, packed into large bags and transported by hand, animal-drawn carts and small trucks to roadside sites where it is collected by trucks and sent to urban wholesale and retail sites. Thus charcoal making is an enterprise for rural populations to supply urban markets. Crop residues and dung are normally used by animal owners as a seasonal supplement to fuelwood.

Shredded biomass residues may be densified by briquetting or pelletizing, typically in screw or piston presses that compress and extrude the biomass. Briquettes and pellets can be good substitutes for coal, lignite and fuelwood because they are renewable and have consistent quality and size, better thermal efficiency, and higher density than loose biomass.

There are briquetting plants in operation in India and Thailand, using a range of secondary residues and with different capacities, but none as yet in other Asian countries. There have been numerous, mostly development agency-funded, briquetting projects in Africa, and most have failed technically and/or commercially. The reasons for failure include deployment of new test units that were not proven technically, selection of very expensive machines that did not make economic sense given the location, low local capacity to fabricate components and provide maintenance, and lack of markets for the briquettes due to uncompetitive cost and low acceptance.

Wood pellets are made of wood waste such as sawdust and grinding dust. Pelletization machines are based on fodder-making technology and produce somewhat lighter and smaller pellets of biomass compared to briquetting. Wood pellets are easy to handle and burn because their shape and characteristics are uniform, transportation efficiency is high and energy density is high. Wood pellets are used as fuel in many countries for cooking and heating applications.

Chips are mainly produced from plantations' waste wood and wood residues (branches and presently even spruce stumps) as a by-product of conventional forestry. They require less processing and are cheaper than pellets. Depending on end use, chips may be produced onsite, or the wood may be transported to the chipper. Chips are commonly used in automated heating systems, and can be used directly in coal-fired power stations or for CHP production.

Charcoal is obtained by heating woody biomass to high temperatures in the absence of oxygen, and has a twice higher calorific value than the original feedstock. It burns without smoke and has a low bulk density, which reduces transport costs. In rural areas in many African countries, charcoal is produced in traditional kilns with efficiencies as low as 10%, and typically sold to urban households while rural households use fuelwood. Hardwoods are the most suitable raw material for charcoal, because softwoods incur possibly high losses during handling/transport. Charcoal from granular materials like coffee shells, sawdust and straw is in powder form and needs to be briquetted with or without a binder. Charcoal is also used in large-scale industries, particularly in Brazil from high-yielding eucalyptus plantations, and in many cases, in conjunction with sustainably produced wood, and also increasingly as a co-fi ring feedstock in oil-based electric power plants. The projected costs for charcoal production from Brazilian eucalyptus plantations are USD_{2005} 5.7 to 9.8/GJ using industrial carbonizing process.

Charcoal in Africa is predominantly produced in inefficient traditional kilns in the informal sector, often illegally. Current production, packaging and transport of charcoal are characterized by low efficiencies and poor handling, leading to losses. Introducing change to this industry requires that it be recognized and legalized, where it is found to be sustainable and not contradictory to environmental protection goals. Once legalized, it would be possible to regulate it and introduce standards addressing fuel quality, packaging and production kiln standards and better enforcement of which tree species should be used to produce charcoal.

Wood Pellet Logistics and Supplies

Wood pellets are one of the most successful bioenergy-based commodities traded internationally. Wood pellets offer several advantages over other solid biomass fuels: they generally have a low moisture content and a relatively high heating value (about 17 GJ/t), which allow long-distance transport by ship without affecting the energy balance. Local transport is carried out by trucks, which sets a feasible upper limit for transportation of 50 km for raw biomass (150 km for pellets) and together with the necessary storage usually represents more than 50% of the final cost. Bulk delivery of pellets is very similar to delivery of home heating oil and is carried out by the lorry driver blowing pellets into the storage space, while a suction pump takes away any dust. Storage solutions include underground tanks, container units, silos or storage within the boiler room. Design of more efficient pellet storage, charging and combustion systems for domestic users is ongoing. International trade by ships to ports that are properly equipped for handling pellets is a major logistical barrier. Freight costs are another barrier very sensitive to international trade demand.

Biochar

Biochar is a charcoal-like substance that's made by burning organic material from agricultural and forestry wastes (also called biomass) in a controlled process called pyrolysis. Although it looks a lot like common charcoal, biochar is produced using a specific process to reduce contamination and safely store carbon. During pyrolysis organic materials, such as wood chips, leaf litter or dead plants, are burned in a container with very little oxygen. As the materials burn, they release little to no contaminating fumes. During the pyrolysis process, the organic material is converted into biochar, a stable form of carbon that can't easily escape into the atmosphere. The energy or heat created during pyrolysis can be captured and used as a form of clean energy. Biochar is by far more efficient at converting carbon into a stable form and is cleaner than other forms of charcoal.

In terms of physical attributes, biochar is black, highly porous, lightweight, fine-grained and has a large surface area. Approximately 70 percent of its composition is carbon. The remaining percentage consists of nitrogen, hydrogen and oxygen among other elements. Biochar's chemical composition varies depending on the feedstocks used to make it and methods used to heat it.

The Concept of Biochar is Rooted in an Ancient Amazonian Practice

Although biochar technology is considered a more recent strategy for carbon sequestration, the practice of adding charred biomass to improve soil quality is not new. This process is modeled after a 2,000-year-old practice in the Amazonian basin, where indigenous people created areas of rich, fertile soils called terra preta (meaning "dark earth").

Whether these soils were intentionally made or are simply a by-product of farming and/or cooking practices is still unclear. But one thing's for sure: The fertility of terra preta is significantly higher than the otherwise famously infertile soils of the Amazon. This explains why plants grown in terra preta soil grow faster, and are more nutrient-dense, than plants grown in neighboring soils. In fact, terra preta soils continue to hold carbon still today.

How to make Biochar: A Closer Look into Biochar Production

Biochar is produced during pyrolysis, a thermal decomposition of biomass in an oxygen-limited environment. The quality of feedstocks, or materials burned, have a direct impact on the quality of the final biochar product. Ideally, clean feedstocks with 10 to 20 percent moisture and high lignin content must be used—some good examples are field residues and woody biomass. Using contaminated feedstocks, including feedstocks from railway embankments or contaminated land, can introduce toxins into the soil, drastically increase soil pH and/or inhibit plants from absorbing minerals. The most common contaminants are heavy metals—including cadmium, copper, chromium, lead, zinc, mercury, nickel and arsenic—and Polycyclic Aromatic Hydrocarbons.

Biochar can be manufactured through low-cost, small-scale production using modified stoves or kilns, or through large-scale, cost-intensive production, which utilizes larger pyrolysis plants and higher amounts of feedstocks. One of the most common ways to make biochar for on-farm use is through pyrolysis using a top-lit updraft biochar machine.

Applications of Biochar in Agriculture: Enhancing Soil and Compost Properties

Soil degradation is a major concern in agriculture globally. To address this burgeoning problem, researchers suggested applying biochar to degraded soils in order to enhance its quality. Some of the ways that biochar may help improve soil qualities include:

- Enhancing soil structure.
- Increasing water retention and aggregation.
- Decreasing acidity.
- Reducing nitrous oxide emissions.
- Improving porosity.

- Regulating nitrogen leaching.
- Improving electrical conductivity.
- Improving microbial properties.

Biochar is also found to be beneficial for composting, since it reduces greenhouse gas emissions and prevents the loss of nutrients in the compost material. It also promotes microbial activity, which in turn accelerates the composting process. Plus, it helps reduce the compost's ammonia losses, bulk density and odor.

First Generation Biofuels

Food crops such as starch, sugar and vegetable oil, as well as animal fats, are the main feedstock sources to first-generation biofuels. These types of feedstock characterize first-generation biofuels, which are defined based on the source the fuel is derived rather than the physical nature of the biofuel itself.

The most popular first-generation biofuels are biodiesel (produced mainly from canola, soybean and barley) and bioethanol (produced mainly from corn, wheat and sugarcane) followed by other types of vegetable oil and biogas. Figures below show that global biofuel production grew steadily from about 23 billion litres per year in 2002 to over 110 billion litres per year in 2012. Growth rates in this market decreased between 2011 and 2012, and the annual production of biofuels in 2015 was roughly equal to 2010 levels.

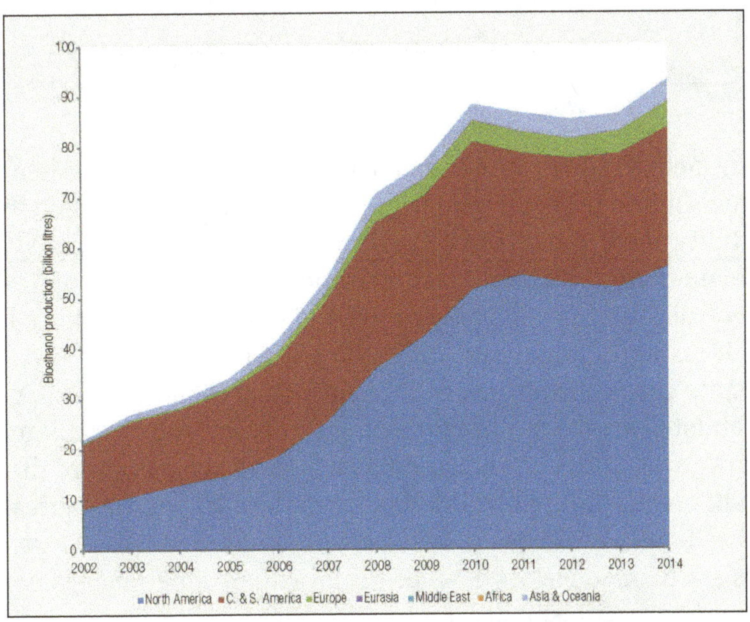

Fig: World bioethanol production, per world region.

Led by Brazil and the United States, ethanol production more than doubled since 2005, reaching 85.6 billion litres in 2010 and 94 billion litres in 2014, and biodiesel production grew from 3.9 billion litres in 2005 to 18.1 billion litres in 2010 and 30 billion litres in 2014.

The Americas produce more than 80 per cent of the global biodiesel and bioethanol production. Ethanol production actually has two major players, namely the United States and Brazil, which cumulatively account for 73 per cent of the global production. Both countries have a history of government support for their biofuel industries, in the same way that the European Union has for its biodiesel industry, which is the largest in the world with 40 per cent of global production.

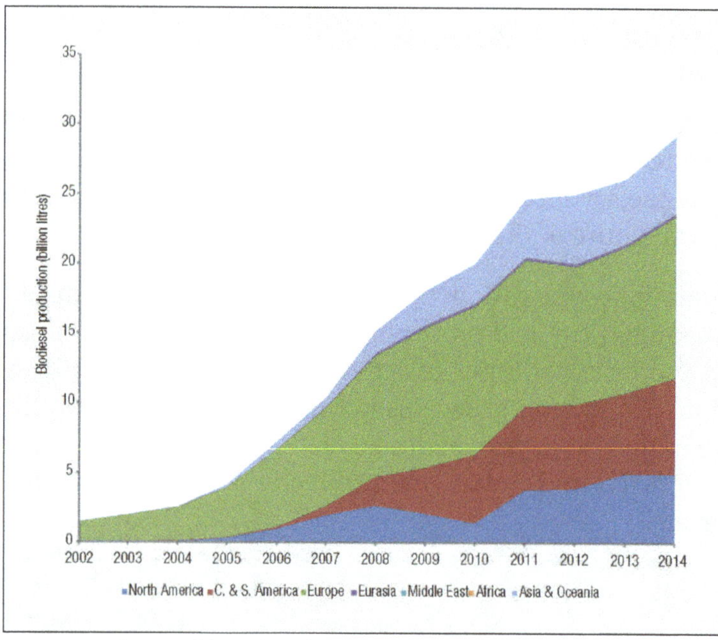

Fig: World biodiesel production, per world region.

There are many factors, which simultaneously constrain and strengthen the growth of the global biofuels market. These include potential negative impacts on food commodities, the availability of natural resources, government subsidies, national commitments to mitigate climate change, oil prices and other political/environmental factors. Biofuel production faces different challenges around the world. Africa suffers from overestimated expectations and agricultural difficulties with some feedstocks, such as Jathropha, but despite these challenges, countries such as Mali, Ghana and Nigeria have established mandates for the use of biofuels. India is working on a biofuel target of 10 per cent and the government has created incentives to production in the form of capital subsidies, tax breaks and public bidding processes. In Latin America, fuel demand is rising and fossil-fuel subsidies are being slowly phased out, at the same time novel biofuel models are being developed, since the Brazilian experience is not replicable in many of its neighbours with smaller land availability.

Table: Characteristics and costs of conventional biofuels.

Biofuel type	Feedstocks	Feedstock characteristics	Estimated production costs US$ cents/litre
Biodiesel	Soy, rapeseed, mustard seed, palm, jatropha, waste vegetable oils and animal fats.	Range of feedstocks with different crop yields per hectare; hence, production costs vary widely among countries. Co-products include high-protein meal.	Soybean oil: 56–72 (Argentina); 100–120 (Global average). Palm oil: 100–130 (Indonesia, Malaysia, and other). Rapeseed oil: 105–130 (EU).
Ethanol	Sugar cane, sugar beets, corn, cassava, sorghum and wheat.	Range of feedstocks with wide yield and cost variations. Co-products include animal feed, heat and power from bagasse residues. Advanced biofuels are starting to become fully commercial and still have higher production costs.	Sugar cane: 82–93 (Brazil). Corn (dry mill): 85–128 (United States).

Biofuels have a potential to overcome environmental challenges by reducing dependence on fossil fuels. Economically, biofuels sustain more than 1.7 million jobs around the world, including 845,000 in Brazil and 282,000 in the United States. However, it is important to note that the emergence of a technological gap between land-intensive first generation and capital-intensive second-generation biofuels can compromise the catch-up strategies implicit on the investments made by the poorer countries.

Second Generation Biofuels

Second generation biofuels are also known as advanced biofuels. What separates them from first generation biofuels the fact that feedstock used in producing second generation biofuels are generally not food crops. The only time the food crops can act as second generation biofuels is if they have already fulfilled their food purpose. For instance, waste vegetable oil is a second generation biofuels because it has already been used and is no longer fit for human consumption. Virgin vegetable oil, however, would be a first generation biofuel.

Because second generation biofuels are derived from different feed stock, Different technology is often used to extract energy from them. This does not mean that second generation biofuels cannot be burned directly as the biomass. In fact, several second generation biofuels, like Switchgrass, are cultivated specifically to act as direct biomass.

Second Generation Extraction Technology

For the most part, second generation feedstock are processed differently than first generation biofuels. This is particularly true of lignocellulose feedstock, which tends to require several processing steps prior to being fermented (a first generation technology) into ethanol.

Thermochemical Conversion

- The first thermochemical route is known as gasification. Gasification is not a new technology and has been used extensively on conventional fossil fuels for a number of years. Second generation gasification technologies have been slightly altered to accommodate the differences in biomass stock. Through gasification, carbon-based materials are converted to carbon monoxide, hydrogen, and carbon dioxide. This process is different from combustion in that oxygen is limited. The gas that result is referred to as synthesis gas or syngas. Syngas is then used to produce energy or heat. Wood, black liquor, brown liquor, and other feedstock are used in this process.
- The second thermochemical route is known as pyrolysis. Pyrolysis also has a long history of use with fossil fuels. Pyrolysis is carried out in the absence of oxygen and often in the presence of an inert gas like halogen. The fuel is generally converted into two products: tars and char. Wood and a number of other energy crops can be used as feedstock to produce bio-oil through pyrolysis.
- A third thermochemical reaction, called torrefaction, is very similar to pyrolysis, but is carried out at lower temperatures. The process tends to yield better fuels for further use in gasification or combustion. Torrefaction is often used to convert biomass feedstock into a form that is more easily transported and stored.

Biochemical Conversion

A number of biological and chemical processes are being adapted for the production of biofuel from second generation feedstock. Fermentation with unique or genetically modified bacteria is particularly popular for second generation feedstock like landfill gas and municipal waste.

Common Second Generation Feedstock

To qualify as a second generation feedstock, a source must not be suitable for human consumption. It is not a requirement that the feedstock be grown on non-agricultural land, but it generally goes without saying that a second generation feedstock should grow on what is known as marginal land. Marginal land is land that cannot be used for "arable" crops, meaning it cannot be used to effectively grow food. The unspoken point here is that second generation feedstock should not require a great deal of water or fertilizer to grow, a fact that has led to disappointment in several second generation crops.

Grasses

A number of grasses like Switchgrass, Myscanthus, Indiangrass, and others have alternatively been placed in the spotlight. The particular grass chosen generally depends on the location as some are more suitable to certain climates. In the United States, Switchgrass is favoured. In Southeast Asia, Myscanthus is the choice.

Advantages of Grasses

- They are perennial and so energy for planting need only be invested once.
- They are fast growing and can usually be harvested a few times per year.
- They have relatively low fertilizer needs.
- They grow on marginal land.
- They work well as direct biomass.
- They have a high net energy yield of about 540%.

Disadvantages of Grasses

- They are not suitable for producing biodiesel.
- They require extensive processing to make into ethanol.
- It may take several years for switch grass to reach harvest density.
- The seeds are weak competitors with weeds. So, even though they grow on marginal land, the early investment in culture is substantial.
- They require moist soil and do not do well in arid climates.

Water demands are the biggest drawback to grasses and the factor that keeps them from becoming more popular as second generation biofuels. Despite this shortcoming, grasses do find a number of uses, particularly in the United States.

Jatropha and other Seed Crops

Seed crops are useful in the production of biodiesel. In the early Part of the 21st century, a plant known as Jatropha became exceedingly popular among biodiesel advocates. The plant was praised for its yield per seed, which could return values as high as 40 percent. When compared to the 15 percent oil found in soybean, Jatropha look to be a miracle crop. Adding to its allure was the misconception that we could be grown on marginal land. As it turns out, oil production drops substantially when Jatropha is grown on marginal land. Interest in Jatropha has waned considerably in recent years. Other, similar seed crops have met with the same fate as Jatropha. Examples include Cammelina, Oil Palm, and rapeseed. In all cases, the initial benefits of the crops were quickly realized to be offset by the need to use crop land to achieve suitable yields.

Waste Vegetable Oil (WVO)

WVO have been used as a fuel for more than a century. In fact, some of the earliest diesel engines ran exclusively on vegetable oil. Waste vegetable oil is considered a second generation biofuels because its utility as a food has been expended. In fact, recycling it for fuel can help to improve its overall environmental impact.

The advantages of WVO are:

- It does not threaten the food chain.
- It is readily available.
- It is easy to convert to biodiesel.
- It can be burned directly in some diesel engines.
- It is low in sulphur.
- There are no associated land use changes.

The disadvantages of WVO are:

- It can decrease engine life if not properly refined.

WVO is probably one of the best sources of biodiesel and, as long as blending is all that is required, can meet much of the demand for biodiesel. Collecting it can be a problem though as it is distributed throughout the world in restaurants and homes.

Municipal Solid Waste

This refers to things like landfill gas, human waste, and grass and yard clippings. All of these sources of energy are, in many cases, simply being allowed to go to waste. Though not as clean as solar and wind, the carbon footprint of these fuels is much less than that of traditionally derived fossil fuels. Municipal solid waste is often used in cogeneration plants, where it is burned to produce both heat and electricity.

Third Generation Biofuels

Microalgae are currently being promoted as an ideal third generation biofuel feedstock because of their rapid growth rate, greenhouse gas fixation ability (net zero emission balance) and high production capacity of lipids (fat). They also do not compete with food or feed crops, and can be grown on non-arable land and saline water. Biofuels are generally referred to solid, liquid or gaseous fuels derived from organic matter. The classification of biofuels is shown in figure below.

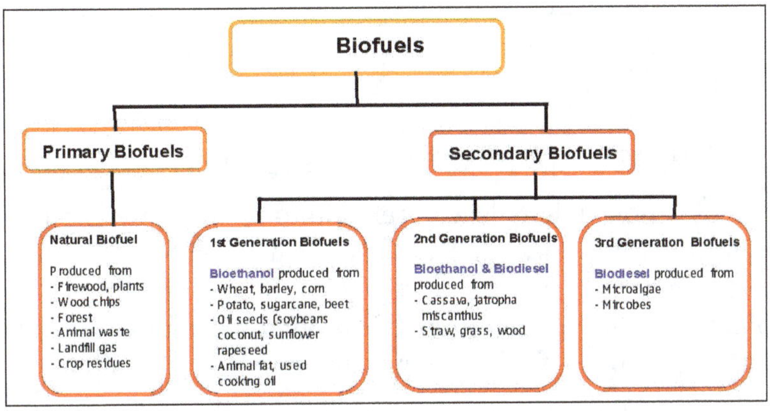

Fig: Biofuel production sources (biomasses).

The first generation biofuels possess notable economic, environmental and political concern as the mass production of biofuel requires more arable agricultural lands resulting in reduced lands for human and animal food production. Moreover, production process of first generation biofuels is also responsible for environmental degradation. Therefore, enthusiasms about first generation biofuels have been demised. As first generation biofuels are not viable, researchers focused on second generation biofuels. Because of the second generation biofuels production process requires expensive and sophisticated technologies, the biofuel production from the second generation is not profitable for commercial production. Therefore, the researchers focused on third generation biofuels. The main component of third generation biofuels is microalgae as shown in Figure. It is currently considered to be a feasible alternative renewable energy resource for biofuel production overcoming the disadvantages of first and second generation biofuels. Microalgae can provide several different types of renewable biofuels. This includes methane, biodiesel and bio-hydrogen. There are many advantages for producing biofuel from algae as microalgae can produce 15 to 300 times more biodiesel than traditional crop on area basis. The harvesting cycle of microalgae is very short and growth rate is very high. Moreover, high quality agricultural land is not required for microalgae biomass production.

Biofuel Production from Microalgae

Microalgae are single-cell microscopic organisms which are naturally found in fresh water and marine environment. There are more than 300,000 species of micro algae, diversity of which is much greater than plants. Microalgae are generally more efficient converters of solar energy comparing to higher plants. In addition, because the cells grow in aqueous suspension, they have more efficient access to water, CO_2, and other nutrients. The current biofuel yields from various biomasses are shown in Table. The table clearly shows huge potential of microalgae compared to other biomasses.

Table: E-Portfolio types based on features and purposes.

Oil Yields	Litre/Hectare/Year	Barrels/Hectare/Year
Soybeans	400	2.5
Sunflower	800	5
Canola	1,600	10
Jathropha	2,000	12
Palm Oil	6,000	36
Microalgae	60,000 – 240,000	360 – 1,500

The oil contents of various microalgae in relation to their dry weight are shown in Table. It is clear that several species of microalgae can have oil contents up to 80% of their dry body weight. Some microalgae can double their biomasses within 24 hours and the shortest doubling time during their growth is around 3.5 hours which makes microalgae an ideal renewable source for biofuel production.

Table: Oil contents of microalgae.

	Name of microalgae	(% dry wt)
1	Botryococcus braunii	25 - 75
2	Chlorella sp.	28 - 32
3	Crypthecodinium cohnii	20
4	Cylindrotheca sp.	16 - 37
5	Dunaliella primolecta	23
6	Isochrysis sp.	25 – 33
7	Monallanthus salina	20
8	Nannochloris sp.	20 - 35
9	Nannochloropsis sp.	31 - 68
10	Neochloris oleoabundans	35 - 54
11	Nitzschia sp.	45 - 47
12	Phaeodactylum tricornutum	20 - 30
13	Schizochytrium sp.	50 - 77
14	Tetraselmis sueica	15 - 23

Microalgal Biomass Production

Producing microalgal biomass is generally more expensive and technologically challenging than growing crops. Photosynthetic growth of microalgae requires light, CO_2, water and inorganic salts. The temperature regime needs to be controlled strictly. For most microalgae growth, the temperature generally remains within 20 °C to 30 °C. In order to reduce the cost, the biofuel production must rely on freely available sunlight, despite daily and seasonal variations in natural light intensities. Growth medium must provide the inorganic elements that constitute the algal cell. Essential elements include nitrogen (N), phosphorus (P), iron (Fe) and in some cases silicon (Si).

Microalgae is grown in various aquatic environments, such as fresh and marine water, municipal waste waters, industrial waste waters and animal waste waters as long as there are adequate amounts of carbon (organic or inorganic), N (urea, ammonium or nitrate), and P as well as other trace elements are present. Sea water supplemented with commercial nitrate and phosphate fertilizers and a few other micronutrients are commonly used for growing marine microalgae. Waste waters are unique in their chemical profile and physical properties as compared with fresh and marine waters. Recent researches indicated the great potential of mass production of algal biomass for biofuel and other applications using wastewaters. However, wastewater based algae cultivation still faces with many uncertainties and challenges including variation of wastewater composition due to source, infrastructure, weather conditions, and pre-treatment methods, improper nutrient ratios (e.g., C/N and N/P), high turbidity due to the presence of pigments and suspended solid particles which affects light transmission, and the presence of competing microflora and toxic compounds, and accumulation of growth inhibiting compounds which is worsened if water is recycled and reused.

There are different ways microalgae can be cultivated. However, two widely used cultivation systems are a) suspended cultures, including open ponds and closed reactors, and b) immobilized cultures, including matrix-immobilized systems and biofilms. The most common large scale production systems in practice are high rate algal ponds or raceway ponds. Raceway ponds are open and shallow with paddle wheel to provide circulation of the algae and nutrients. Raceways are relatively inexpensive to build and operate, but often suffer low productivity for various reasons. Tubular photobioreactors are the only type of closed systems used at large scale production of algae.

The photoreactor system can be sub-classified as: a) vertical photoreactor, b) flat or horizontal photoreactor, and c) helical photoreactor. The helical photoreactor is considered the easiest to scale up production. Compared to open ponds, tubular photobioreactors can give better pH and temperature control, better protection against culture contamination, better mixing, less evaporative loss and higher cell densities. However, each system has relative advantages and disadvantages. One of the significant challenges of using raceways and tubular photobioreactors is biomass recovery. This challenge has been mitigated to an extent by immobilized cultures or attached algal processes. Algal biofilms could play a large role in overcoming the major challenges to production and harvesting of microalgae. If enough surface area is provided, algae biofilm growth can be more than suspended growth.

Algae Harvesting Methods

Various methods are currently used for harvesting algae, which includes chemical based, mechanical based, biological based and to a lesser extent, electrical based operations However, various combinations or sequence of these methods are also commonly in use. The cell size of algae is very small. Therefore, chemical flocculation is often performed as a pre-treatment to increase the particle size of algae before using another method such as flotation to harvest the algae. In mechanical based process, centrifugation process, which is the most reliable and rapid method, is used for recovering suspended algae. In electrical based method, negative charge properties of algal cells are used for separating the cells. These cells can be concentrated by the movement in an electric field.

Biofuels Production Processes from Microalgae

A number of ways the microalgae biomass can be converted into energy sources which includes: a) biochemical conversion, b) chemical reaction, c) direct combustion, and d) thermochemical conversion. Figure illustrates a schematic of biodiesel and bioethanol production processes using microalgae feedstock. Depending on the microalgae species, in addition to biofuel production, other compounds may also be extracted, with valuable applications in different industrial sectors. The production of biofuel is a complex process. A schematic of biofuel production processes from microalgae is shown in Figure. However, these processes are complex, technologically challenging and economically expensive.

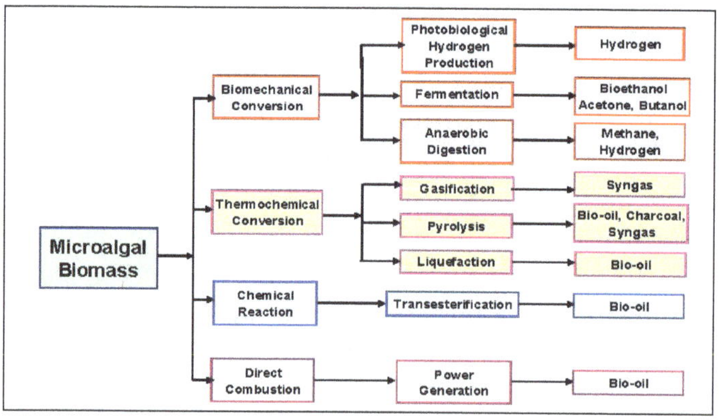

Biofuel production processes from microalgae biomass.

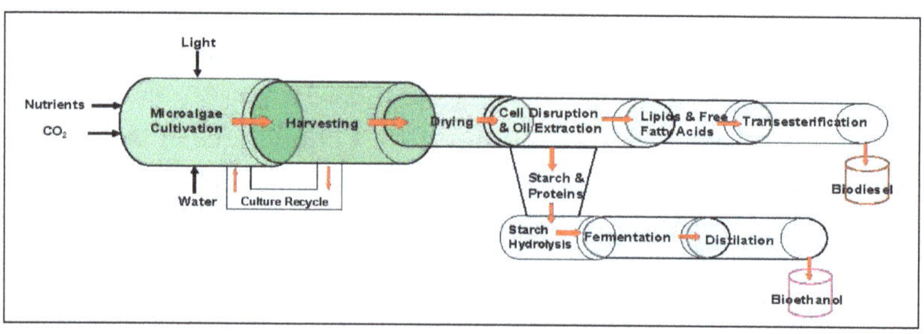

Biodiesel and Bioethanol production processes from microalgae.

Fourth Generation Biofuels

Fourth generation biofuels take advantage of synthetic biology of algae and cyanobacteria, which is a young but strongly evolving research field. Synthetic biology comprises the design and construction of new biological parts, devices and systems, and the re-design of existing, natural biological systems for useful purposes. It is becoming possible to design a photosynthetic/non-photosynthetic chassis, either natural or synthetic, to produce high-quality biofuels with high PFCE. For the first, second and third generation biofuels, and the raw material is either biomass or waste, both being results of "yesterday's photosynthesis" (yet not from fossil resources). While these biofuels often are very useful in a certain region or country, they are always limited by the availability of the corresponding organic raw material, i.e. the biomass, which limits their application on a global scale.

Fourth generation biofuels will be based on raw materials that are essentially inexhaustible, cheap and widely available. Photosynthetic water splitting (water oxidation) into its constituents by solar energy can become a large contributor to fuel production on a global scale, both by artificial photosynthesis and by direct solar biofuel production technologies. Not only the production of hydrogen but also the production of reduced carbon-based biofuels is possible by concomitant enhanced fixation of atmospheric CO_2 and innovative design of synthetic metabolic pathways for fuel production. The generation of "designer bacteria" with new useful properties requires revolutionary scientific breakthroughs in several areas of fundamental research.

The European Union bioeconomy strategy highlights the importance of discovery research for the establishment of functional bioeconomy, and synthetic biology is foreseen as a key enabling technology for successful realisation of bioeconomy in replacement of fossil fuels. Synthetic biology will have the capacity to make bioeconomy much broader by providing means to produce numerous different biological compounds, including an array of various biofuels. There is also a worry about the decrease in EU's biocapacity. If no actions will be taken, it is forecasted that consumption of bioresources will exceed their renewal capacity by 2030. Thus, it is considered extremely important to produce biofuels using minimal raw material resources in their production. Fourth generation biofuels are produced by:

- Designer photosynthetic microorganisms to produce photobiological solar fuels.
- Combining photovoltaics and microbial fuel production (electrobiofuels).
- Synthetic cell factories or synthetic organelles specifically tailored for production of desired high-value chemicals (production currently based on fossil fuels) and biofuels.

Designer Microorganisms in Production of Solar Biofuels

Key improvements in photon-to-fuel conversion efficiency (PFCE) as well as in the quality of the fuel will be based on generation of designer microorganisms. They produce a selected photobiological solar fuel, which is a non-fossil fuel made by direct solar energy conversion in photosynthetic microorganisms (algae or cyanobacteria), and are engineered, when necessary, to secrete the fuel). Photobiological solar fuel is made in photosynthetic cell from solar energy using only water, or water and CO_2 as raw materials, depending whether the produced fuel is hydrogen- or carbon-based fuel, respectively. Knowledge based on powerful biochemical and biophysical insights in natural photosynthetic light harvesting, water splitting, electron transfer, hydrogenases and carbon metabolism is critical for development of direct photobiological solar fuels. Microorganisms will be optimised for fuel production with synthetic biology approaches, metabolic engineering and organism design based on knowledge acquired by genomic research, molecular systems biology and extensive modelling research.

Most advanced research so far made towards direct photobiological solar fuel production is based on research with unicellular algae and cyanobacteria. Cyanobacteria are suitable as photosynthetic chassis for their well-developed genetic transformation technologies as well as for extensive knowledge on their light harvesting and electron transfer processes, on their metabolome and advanced modelling research of the entire cell. Cyanobacteria have been genetically engineered to produce various fuels and chemicals (e.g. H_2, ethanol, isobutanol, isoprene, lactic acid). Introduction of various fermentative metabolism pathways to cyanobacteria cell by synthetic biology approaches has made it possible to produce biofuels directly from solar energy and Calvin–Benson cycle intermediates. Furthermore, the efficient secretion of product from the cells will increase the production capacity.

Breakthroughs in direct photobiological solar fuel production call for further research in design of light-harvesting systems, modelling and simulation of biological reactions and systems as well as in development of synthetic biology tools and production systems. Synthetic biology tools comprise the analysis, design and synthesis of biological systems based on well-defined and ideally orthogonal functional modules. Although the first results on direct photobiological solar fuels have proven their potential in fuel production, this field is still at the level of basic research. During the coming 10–20 years, it is expected that various photobiological solar fuels are gradually entering into the market.

Electrobiofuels

It is possible, through synthetic biology approaches, to establish new-to-nature hybrid production organisms that use renewable electricity and carbon sources for production of commodity chemicals and biofuels. A newly emerging field, MES relies on the capability of certain microbes for direct electron capture from electrodes (e.g. from solar cells or any renewable electricity source) to assimilation of the reducing equivalents into metabolism, along with CO_2 utilisation. Through MES, energy from solar cells can be turned into storable energy sources called electrofuels. Electrofuels allow renewables from all sources to be stored conveniently and efficiently as a liquid fuel. Bacteria, suitable for MES, transport electrons through the outer membrane via specific protein-embedded electron conduits. Direct electron transfer was estimated to be the most efficient mechanism for electron transfer from electrodes to microorganisms.

Production of fuel butanol has already been demonstrated with the engineered MES organism Clostridium ljungdahlii. This technology may provide in the future an efficient way for storage of solar energy in the form of liquid fuel.

Recently, researchers have demonstrated a novel and scalable integrative bioelectrochemical production system for isopropanol. The solar water-splitting catalyst, based on earth-abundant metals, was used to provide energy for growth of a bacterium Ralstonia eutropha. In metabolically engineered Ralstonia, the energy from water splitting could be diverted for production of isopropanol. Authors claim the highest bioelectrochemical fuel yield reported so far. Indeed, they provide an important proof-of-principle demonstration that inorganic and biological materials can be combined to achieve higher PFCE than either system can provide independently.

From Current Biorefineries to Synthetic Factories in Production of Solar Biofuels

The design of synthetic factories and cell organelles for enhanced biofuel production is a future technology and currently just entering into intense developmental phase at the basic level of biofuel research. Likewise, the synthetic biology technology itself is still young and only a few truly synthetic examples have been realized by now. For optimal solutions, one needs to go beyond the known possibilities offered by biology by speeding up evolution and screening a massive number of artificial biological combinations. Thus, automation, microfabrication and measurement technologies are an integral part of a successful synthetic biology platform. Most importantly, engineering principles should be used to guide modelling, design and standardisation of the synthetic biological systems themselves and the entire development process so that optimisation can progress from the current trial-and-error situation to the design of truly programmable biofuel production systems.

Synthetic biology, after reaching maturity as a technology, will give tools and concepts that will make biology fully engineerable, and thereby make it possible to take full advantage of the diversity, functionality and specificity that biology can offer. Such research and technology development programmes should be established for construction of synthetic factories and organelles towards efficient biofuel production.

2

Diverse Varieties of Biofuels

Biofuels are classified into various types. A few such types are biogas, bio alcohol fuels, aviation biofuel, syngas and biodiesel. Biogas is produced anaerobically by organic matter and bio alcohol fuels are generally produced using aliphatic alcohols such as methanol, ethanol, propanol and butanol. All these diverse types of biofuels have been carefully analyzed in this chapter.

Biogas

Biogas is a type of biofuel that is naturally produced from the decomposition of organic waste. When organic matter, such as food scraps and animal waste, break down in an anaerobic environment (an environment absent of oxygen) they release a blend of gases, primarily methane and carbon dioxide. Because this decomposition happens in an anaerobic environment, the process of producing biogas is also known as anaerobic digestion.

Anaerobic digestion is a natural form of waste-to-energy that uses the process of fermentation to breakdown organic matter. Animal manure, food scraps, wastewater, and sewage are all examples of organic matter that can produce biogas by anaerobic digestion. Due to the high content of methane in biogas (typically 50-75%) biogas is flammable, and therefore produces a deep blue flame, and can be used as an energy source.

The Ecology of Biogas

Biogas is known as an environmentally-friendly energy source because it alleviates two major environmental problems simultaneously:

- The global waste epidemic that releases dangerous levels of methane gas every day.
- The reliance on fossil fuel energy to meet global energy demand.

By converting organic waste into energy, biogas is utilizing nature's elegant tendency to recycle substances into productive resources. Biogas generation recovers waste materials that would otherwise pollute landfills; prevents the use of toxic chemicals in sewage treatment plants, and saves money, energy, and material by treating waste on-site. Moreover, biogas usage does not require fossil fuel extraction to produce energy.

Instead, biogas takes a problematic gas, and converts it into a much safer form. More specifically, the methane content present in decomposing waste is converted into carbon dioxide. Methane gas has approximately 20 to 30 times the heat-trapping capabilities of carbon dioxide. This means that when a rotting loaf of bread converts into biogas, the loaf's environmental impact will be about 10 times less potent than if it was left to rot in a landfill.

Biogas Digesters

As opposed to letting methane gas release to the atmosphere, biogas digesters are the systems that process waste into biogas, and then channel that biogas so that the energy can be productively used. There are several types of biogas systems and plants that have been designed to make efficient use of biogas. While each model differs depending on input, output, size, and type, the biological process that converts organic waste into biogas is uniform. Biogas digesters receive organic matter, which decompose in a digestion chamber. The digestion chamber is fully submerged in water, making it an anaerobic (oxygen-free) environment. The anaerobic environment allows for microorganisms to break down the organic material, and convert it into biogas.

All-Natural Fertilizer

Because the organic material decomposes in a liquid environment, nutrients present in the waste dissolve into the water, and create a nutrient-rich sludge, typically used as fertilizer for plants. This fertilizer output is generated on a daily basis, and therefore is a highly productive by-product of anaerobic digestion.

Biological Breakdown

To produce biogas, organic matter ferments with the help of bacterial communities. Four stages of fermentation move the organic material from their initial composition into their biogas state:

- The first stage of the digestion process is the hydrolysis stage. In the hydrolysis stage insoluble organic polymers (such as carbohydrates) are broken down, making it accessible to the next stage of bacteria called acidogenic bacteria.
- The acideogenic bacteria convert sugars and amino acids into carbon dioxide, hydrogen, ammonia, and organic acids.
- At the third stage the acetogenic bacteria convert the organic acids into acetic acid, hydrogen, ammonia, and carbon dioxide, allowing for the final stage- the methanogens.
- The methanogens convert these final components into methane and carbon dioxide- which can then be used as a flammable, green energy.

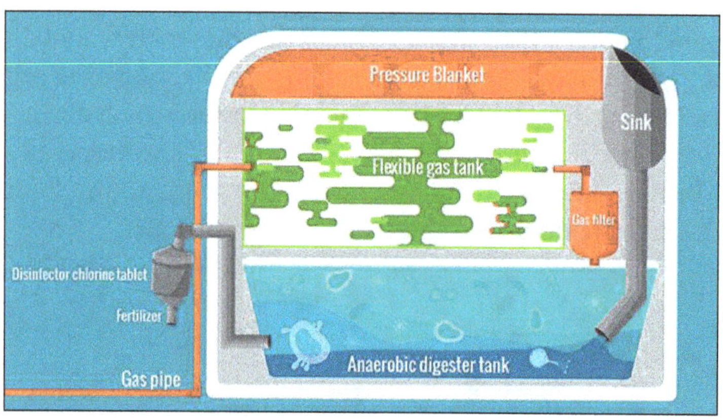

This anaerobic process of decomposition (or fermentation) of organic matter happens all around us in nature, and has been happening for a very long time. In fact, the bacteria that break down organic material into biogas are some of the oldest multi-celled organisms on the planet. Human use of biogas, of course doesn't go that far back, however, some anecdotal evidence traces the first uses of biogas to the Assyrians in the 10th century and the Persians in the 16th century. More recently, the 20th century has brought about a renaissance of both industrial and small-scale biogas systems.

In the 18th century it became clear to Flemish chemist Jan Baptise van Helmont that decomposing organic matter produced a combustible gas. Soon after, John Dalton and

Humphrey Davy clarified that this flammable gas was methane. The first major anaerobic digestion plant dates back to 1859 in Bombay. Not long after, in 1898, the UK used anaerobic digestion to convert sewage into biogas, which was then used to light street lamps. For the next century, anaerobic digestion was primarily used as a means to treat municipal wastewater. When the price of fossil fuels rose in the 1970's industrial anaerobic digestion plants increased in popularity and efficiency.

Both India and China began developing small-scale biogas digesters for farmers around the 1960's. The goal was to decrease energy poverty in rural areas, and make cleaner cooking fuels more accessible in remote areas. Close to one third of the global population still uses firewood and other biomass for energy, causing devastating health and environmental problems.

The popular model is known as the floating drum digester, and China's preferred biogas model is called the fixed dome digester.

Since then, family-sized biogas units are gaining more attention and popularity as both a means of reducing household waste and as a means of providing clean renewable energy to families throughout the world. In the past 15 years, countries around the globe are adopting biogas programs to make both household biogas systems and larger anaerobic digestion plants accessible, efficient, and convenient. As landfills get illegally overloaded, and as the release of methane poses more worrying problems, the benefits of using biogas systems to convert waste into energy are increasingly more relevant and important.

Many Uses of Biogas

Biogas can be produced with various types of organic matter, and therefore there are several types of models for biogas digesters. Some industrial systems are designed to treat: municipal wastewater, industrial wastewater, municipal solid waste, and agricultural waste.

Small-scale systems are typically used for digesting animal waste. And newer family-size systems are designed to digest food waste. The resulting biogas can be used in several ways including: gas, electricity, heat, and transportation fuels.

For example, in Sweden hundreds of cars and buses run on refined biogas. The biogas in Sweden is produced primarily from sewage treatment plants and landfills.

Another example of the diversified uses of biogas is the First Milk plant. One of the UK's biggest cheese makers is building an anaerobic digestion plant that will process dairy residues and convert into bio-methane for the gas grid.

Small-Scale Biogas Systems

Small-scale or family-size biogas digesters are most frequently found in China. However, the demand for such units is growing rapidly throughout the world thanks to more advanced and convenient technologies. As the modern world is producing more and more waste, individuals are eager to find ecologic ways to treat their trash.

Traditional systems typically found in China focuses on animal waste. Due to a lack of energy in rural areas combined with a surplus of animal manure, biogas digesters are very popular, useful, and even life-changing. In many developing countries, biogas digesters are even subsidized and advocated by the government and local ministries, who see the variety of benefits produced from using biogas. In addition to having a clean renewable energy provide gas in the kitchen, many families make extensive use of the fertilizer by-product that biogas digesters provide.

In African countries, some biogas users even turn a profit by selling the bio-slurry by-product produced by biogas systems. This bio-slurry is different from the liquid fertilizer that is produced daily. Bio-slurry refers to the most decomposed stage of the organic matter, after it has been broken down in the system. Bio-slurry sinks to the bottom of the biogas system, and with the help of modern units is easily emptied out once accrued (usually an annual process). This bio-slurry is in fact a nutrient-dense sludge that provides lots of benefits to soil, and can increase productivity of vegetable gardens.

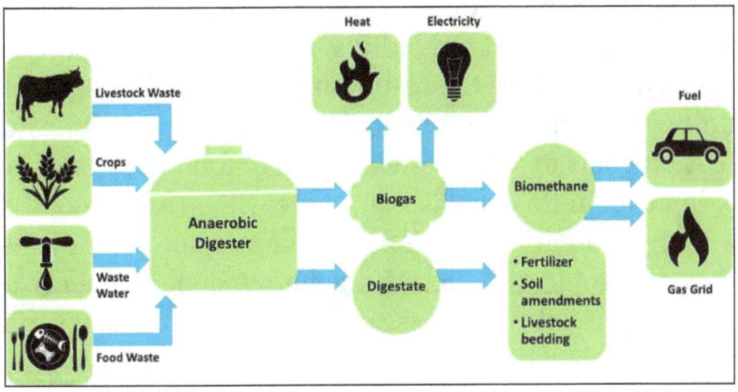

Fig: Anaerobic digestion process.

The Benefits of Biogas

Stored biogas can provide a clean, renewable, and reliable source of baseload power in place of coal or natural gas. Baseload power is consistently produced to meet minimum power demands; renewable baseload power can complement more intermittent renewables. Similar to natural gas, biogas can also be used as a source of peak power that can be rapidly ramped up. Using stored biogas limits the amount of methane released into the atmosphere and reduces dependence on fossil fuels. The reduction of methane emissions derived from tapping all the potential biogas in the United States would be equal to the annual emissions of 800,000 to 11 million passenger vehicles. Based on a waste-to-wheels assessment, compressed natural gas derived from biogas reduces greenhouse gas emissions by up to 91 percent relative to petroleum gasoline.

In addition to climate benefits, anaerobic digestion can lower costs associated with waste remediation as well as benefit local economies. Building the 13,500 potential biogas systems in the United States could add over 335,000 temporary construction jobs and 23,000 permanent jobs. Anaerobic digestion also reduces odors, pathogens, and the risk of water pollution from livestock waste. Digestate, the material remaining after the digestion process, can be used or sold as fertilizer, reducing the need for chemical fertilizers. Digestate also can provide additional revenue when sold as livestock bedding or soil amendments.

Biogas Feedstocks

Food Waste

Around 30 percent of the global food supply is lost or wasted each year. In 2010 alone, the United States produced roughly 133 billion pounds (66.5 million tons) of food waste, primarily from the residential and commercial food sectors. To address this waste, EPA's Food Recovery Hierarchy prioritizes source reduction first, then using extra food to address hunger; animal feed or energy production are a lower priority. Food should be sent to landfills as a last resort. Unfortunately, food waste makes up 21 percent of U.S. landfills, with only 5 percent of food waste being recycled into soil improver or fertilizer. Most of this waste is sent to landfills, where it produces methane as it breaks down. While landfills may capture the resultant biogas, landfilling organic wastes provides no opportunity to recycle the nutrients from the source organic material. In 2015, the EPA and USDA set goals to reduce the amount of food waste sent to landfills by 50 percent by 2030. But even if this goal is met, there will be excess food that will need to be recycled. The energy potential is significant. As just one example, with 100 tons of food waste per day, anaerobic digestion can generate enough energy to power 800 to 1,400 homes each year. Fat, oil, and grease collected from the food service industry can also be added to an anaerobic digester to increase biogas production.

Landfill Gas

Landfills are the third largest source of human-related methane emissions in the United States. Landfills contain the same anaerobic bacteria present in a digester that break down organic materials to produce biogas, in this case landfill gas (LFG). Instead of allowing LFG to escape into the atmosphere, it can be collected and used as energy. Currently, LFG projects throughout the United States generate about 17 billion kilowatt-hours of electricity and deliver 98 billion cubic feet of LFG to natural gas pipelines or directly to end-users each year. For reference, the average U.S. home in 2015 used about 10,812 kilowatt-hours of electricity per year.

Livestock Waste

A 1,000-pound dairy cow produces an average of 80 pounds of manure each day. This manure is often stored in holding tanks before being applied to fields. Not only does the manure produce methane as it decomposes, it may contribute to excess nutrients in waterways. In 2015, livestock manure management contributed about 10 percent of all methane emissions in the United States, yet only 3 percent of livestock waste is recycled by anaerobic digesters. When livestock manure is used to produce biogas, anaerobic digestion can reduce greenhouse gas emissions, reduce odors, and reduce up to 99 percent of manure pathogens. There is the potential for 8,241 livestock biogas systems, which could together generate over 13 million megawatt-hours of energy each year.

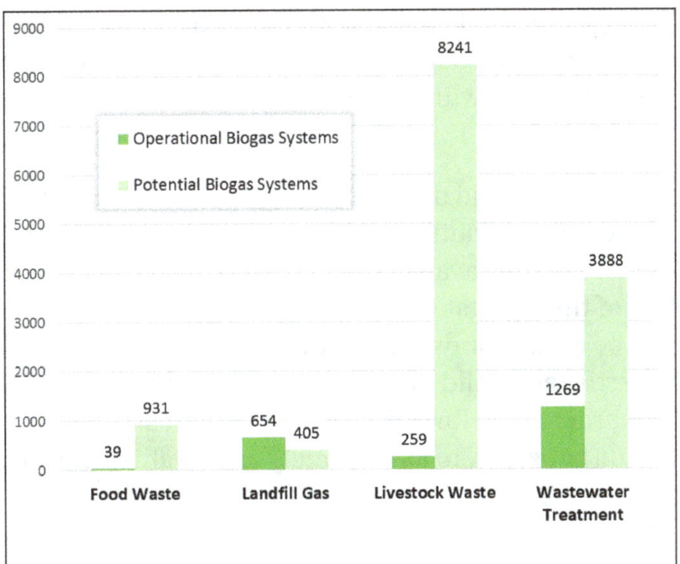

Fig: Current number of operational and potential biogas systems in the United States by feedstock.

Wastewater Treatment

Many wastewater treatment plants (WWTP) already have on-site anaerobic digesters to treat sewage sludge, the solids separated during the treatment process. However, many WWTP do not have the equipment to use the biogas they produce, and flare it instead. Of the 1,269 wastewater treatment plants using an anaerobic digester, only around 860 use their biogas. If all the facilities that currently use anaerobic digestion—treating over 5 million gallons each day—were to install an energy recovery facility, the United States could reduce annual carbon dioxide emissions by 2.3 million metric tons—equal to the annual emissions from 430,000 passenger vehicles.

Crop Residues

Crop residues can include stalks, straw, and plant trimmings. Some residues are left on the field to retain soil organic content and moisture as well as prevent erosion. However, higher crop yields have increased amounts of residues and removing a portion of these can be sustainable. Sustainable harvest rates vary depending on the crop grown, soil type, and climate factors. Taking into account sustainable harvest rates, there are currently around 104 million tons of crop residues available at a price of $60 per dry ton. Crop residues are usually co-digested with other organic waste because their high lignin content makes them difficult to break down.

Biogas End Uses

Raw Biogas and Digestate

With little to no processing, biogas can be burned on-site to heat buildings and power boilers or even the digester itself. Biogas can be used for combined heat and power (CHP) operations, or biogas can simply be turned into electricity using a combustion engine, fuel cell, or gas turbine, with the resulting electricity being used on-site or sold onto the electric grid.

Digestate is the nutrient-rich solid or liquid material remaining after the digestion process; it contains all the recycled nutrients that were present in the original organic material but in a form more readily available for plants and soil building. The composition and nutrient content of the digestate will depend on the feedstock added to the digester. Liquid digestate can be easily spray-applied to farms as fertilizer, reducing the need to purchase synthetic fertilizers. Solid digestate can be used as livestock bedding or composted with minimal processing. Recently, the biogas industry has taken steps to create a digestate certification program, to assure safety and quality control of digestate.

Renewable Natural Gas

Renewable natural gas (RNG), or biomethane, is biogas that has been refined to remove carbon dioxide, water vapor, and other trace gases so that it meets natural gas industry standards. RNG can be injected into the existing natural gas grid (including pipelines) and used interchangeably with conventional natural gas. Natural gas (conventional and renewable) provides 26 percent of U.S. electricity, and 40 percent of natural gas is used to produce electricity. The remainder of natural gas is used for commercial purposes (heating and cooking) and for industrial ones. RNG has the potential to replace up to 10 percent of the natural gas used in the United States.

Compressed Natural Gas and Liquefied Natural Gas

Like conventional natural gas, RNG can be used as a vehicle fuel after it is converted to compressed natural gas (CNG) or liquefied natural gas (LNG). The fuel economy of

CNG-powered vehicles is comparable to that of conventional gasoline vehicles and can be used in light- to heavy-duty vehicles. LNG is not as widely used as CNG because it is expensive to both produce and store, though its higher density makes LNG a better fuel for heavy-duty vehicles that travel long distances. To make the most of investments in fueling infrastructure, CNG and LNG are best suited for fleet vehicles that return to a base for refueling. It is estimated RNG could replace five percent of the natural gas used to produce electricity and 56 percent of the natural gas used to produce vehicle fuel.

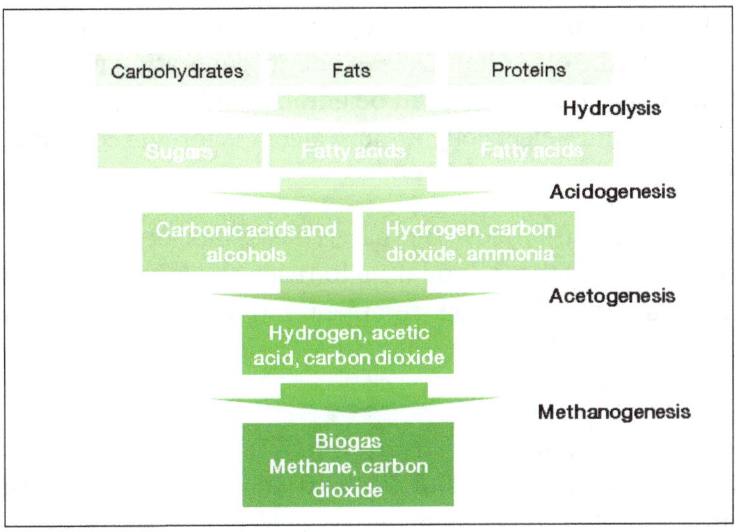

Bio Alcohol Fuels

Alcohol fuels were originally regarded as an alternative energy sources for petroleum oil to realize energy independence during oil crisis of the 1970s. A brief look into the history of bio-ethanol shows Ford Motor Company's development of ethanol-fueled car in 1899, which was terminated by low-priced gasoline then. Oil crisis of the 1970s revived similar interest in the form of gasohol by mixing ethanol, which was developed and commercialized mainly in Brazil.

The first starting point on alcohol fuels in the 1970s tells the basic background at that time. It is prompted by concerns about reliance on foreign sources of oil and a desire to support domestic agriculture. In the United States, in particular, E10 gasohol was implemented during the oil crisis of 1970s to reduce petroleum oil dependence and simultaneously to utilize surplus farm crops. At present, E15 product with 15% ethanol content is distributed for consumer market.

In the twenty-first century, alcohol fuels are again becoming a frequent keyword for clean fuel utilization in connection with mitigation of climate change and clean fuel technology suitable for less-used local energy sources. As a matter of fact, demand for alcohol fuels is mainly derived from socioeconomic and political motivations rather than from consumer conscious reasons and economic viability.

The centralized energy system that emphasizes cost-effectiveness had diminished the key driving force for technological advances in alcohol fuels. Petroleum-based liquid fuel has dominated the transportation area till now. Also, low petroleum oil cost lessened the motivation for further technology development for alcohol fuels. Global oil shock of the 1970s are not expected to break out again within the foreseeable future, and the prospect for alcohol fuel as a remedy to soaring petroleum price is not a plausible picture either. On the other hand, clean energy generation policy by utilization of locally acquired biomass or sea algae will be emphasized to replace local consumption of liquid fuel and to produce electricity or pure alcohols for fuel cells or other means, as a rather cleaner way.

For the future energy sources, renewable-based energy society must be the final goal to reach, but unfortunately it takes a long time to reach the economics and technological easiness to be a common practice, which appears to take at least one or two decades. In order to bring the technology in earlier time, there exist many hurdles and require efforts in scientific and societal side.

All in all, future energy generation direction had been solidly established as "to be clean, renewable, and sustainable," but low petroleum cost lessened the necessity of alternative clean energy source development, e.g., alcohol fuels.

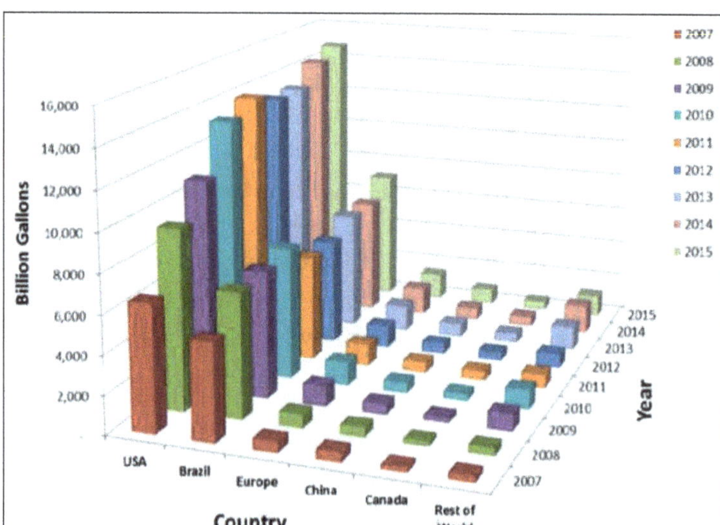

Fig: Worldwide ethanol production by country and year 2007–2015.

Recently, global warming is becoming a central social issue attracting worldwide attention and provides a kind of consensus that society should be changed to deal with alleviating the prime causes of CO_2 evolution in addition to pollution-related issues such as fine dust. The utilization of alcohol fuels reduces carbon dioxide contents in the atmosphere, thus significantly alleviating global warming potential.

Alcohol fuels have been known as a good replacement of fossil-based liquid fuels. Brazil and the United States consume alcohol fuels in the most significant proportions, and such trend will not easily change. In particular, bio-ethanol is well known for its use in Brazil as a gasoline supplement.

Properties of Alcohol Fuels

Alcohols and ethers can replace gasoline and oil. Table below exhibits properties of n-butanol, ethanol, and gasoline for comparison. In Table below, RON, MON, and RVP values for butanol and ethanol are meant for gasoline blend fuels.

Table: Properties of n-butanol and ethanol with gasoline.

Item	n-Butanol	Ethanol	Gasoline
Specific gravity @ 60°F	0.814	0.794	0.720–0.775
Heating value, MJ/L	26.9–27.0	21.1–21.7	32.2–32.9
Research octane number (RON)	94	106–130	95
Motor octane number (MON)	80–81	89–103	85
Reid vapor pressure (RVP) of 5 and 10% Alcohol/ gasoline blends, psi	6.4/6.4	31/20	<7.8/15 (summer/ winter)
Oxygen, wt%	21.6	34.7	<2.7
Water solubility at 25°C, %	9.1	100.0	<0.01

Table below contains a more wide range of properties of alcohols and ethers compared to gasoline and fusel oil. Fusel oil or fusel alcohol is defined as a mixture of several alcohols produced as a by-product of alcoholic fermentation.

Table: Detailed properties of alcohols, ethers, and related fuels.

Item\fuel	Gasoline	Butanol	Methanol	Ethanol	MTBE	DME	Fusel oil
Chemical formula	$C_{5-10}H_{12-22}$	$C_4H_{10}O$	CH_3OH	C_2H_5OH	$C_5H_{12}O$	CH_3-O-CH_3	$C_5H_{12}O$
Molecular weight	106.22	74.12	32.04	46.7	88.15	46.07	76.42
Carbon, mass%	87.5	64.91	37.5	52.2	66.1	52.2	54.8
Hydrogen, mass%	12.5	13.49	—	34.7	13.7	13	15
Oxygen, mass%	0	21.6	49.93	34.7	18.2	34.8	30.32
Density, g/ml	0.737	0.810	0.792	0.785	0.74	0.661	0.847
Boiling temperature, °C	27–225	117.25	78	78.25	52.2	−25.1	53.4–54.4
Reid vapor pressure, Kpa	53–60	18.6	32.4	17	54.47	—	—

Research octane no.	90–100	98	108.7	108.6–110	118	—	106.85
Motor octane no.	82–90	78	86.6	92	102	—	103.72
Low heating value, MJ/kg	44.0	33.2	20.1	26.9	34.9	28.8	29.536
Freezing point, °C	-40	—	-97.5	-114	-108	—	-52
Viscosity, mm²/s	0.5–0.6	—	0.596	1.2–1.5	0.35	—	0.61
Flash point, °C	-45 to -13	—	11	12–20	-25.5	—	—
Auto-ignition temperature, °C	257	385	423	425	435	253	41.6

Table: Alcohol solubility in water in mol/100 g of H_2O.

Alcohol name	Chemical formula	Water solubility
Methanol	CH_3OH	Miscible
Ethanol	C_2H_5OH	Miscible
Propanol	C_3H_7OH	Miscible
Butanol	C_4H_9OH	0.11
Pentanol	$C_5H_{11}OH$	0.03
Hexanol	$C_6H_{13}OH$	0.0058
Heptanol	$C_7H_{15}OH$	0.0008

In general, alcohols contain higher values than gasoline in oxygen content, octane number, and autoignition/flash point temperatures, while freezing point temperature is lower.

Tetraethyllead has been banned for use as an additive to improve octane number of gasoline fuel. Methyl tertiary-butyl ether (MTBE) and alcohols are thus used as alternative additives to gasoline, but MTBE has also been banned after the 2000s, and alcohols have become useful additives to increase octane number of gasoline.

Water solubility of alcohols is an important property when alcohols are being used as fuel. Gasoline has a water solubility value of less than 0.01, whereas ethanol exhibits a full miscibility as 100. When alcohols contain a high solubility in water, spill or leakage of the mixed alcohol fuels can cause polluting the underground water.

Methanol, ethanol, and propanol are completely miscible in water, which means that they dissolve in water in any amount. Both methanol and ethanol dissolve readily in water, are fortunately biodegradable, and do not bioaccumulate. They are not rated as toxic to aquatic organisms.

Starting with the four-carbon alcohol (butanol), solubility is starting to decrease, and from the seven-carbon length heptanol, alcohols are practically immiscible in waters. This is one of the backgrounds for the development of butanol as another alcohol fuel.

Other important properties of alcohol fuels reside in its inherent swelling of plastics and corroding power for metals. These properties ask modification in the existing infrastructure of automobiles and other appliances.

Ethanol

Ethanol is a clear, colorless, toxic liquid and has a characteristic odor. Ethanol is not classified as toxic to humans. Ethanol has a higher octane number than gasoline, providing premium blending properties as a liquid fuel. Ethanol contains less energy per volume than gasoline, and denatured ethanol (98% ethanol) contains about 30% less energy than gasoline per volume. Since ethanol contains oxygen, using it as a gasoline additive results in up to 25% fewer carbon monoxide emissions than conventional gasoline.

Ethanol is soluble in polar and nonpolar solvents and has a clearly higher vapor pressure than gasoline and an oxygen content of approximately 35%. Ethanol itself is a good solvent and can be mixed with water in unlimited quantities. Because ethanol is a short-lived compound in surface water and subsurface aquifer, substantially limiting the risk to aquatic organisms, environmental problem is minimal even when it is spilled. Ethanol degrades quickly in the natural environment, and the biodegradation is rapid in soil, groundwater, and surface water, with predicted half-lives ranging from several hours to 10 days.

Methanol

Methanol, or wood alcohol, is a colorless, odorless, toxic liquid and is the simplest form (CH_3OH) among alcohols. Methanol is corrosive to some materials. Methanol can be produced from several sources: synthetic gas (syngas), formic acid, formaldehyde, and methane. Methanol is classed as toxic so it requires additional considerations during usage to limit inhalation exposure and skin contact.

Methanol is hygroscopic, meaning that it will absorb water vapor directly from the atmosphere. Because absorbed water dilutes the fuel value of the methanol and may cause phase separation of methanol-gasoline blends, containers of methanol fuels must be kept tightly sealed.

Butanol

Butanol has higher energy densities and could be distributed in the existing infrastructure. The use of ethanol as an additive to gasoline to increase octane number has downside effects such as corrosion of metal component and vapor lock. Such troubleshooting can be remedied by modification of engine and fuel system, but addition of alcohols with high carbon number such as bio-butanol enables utilization in existing system without rendering any change.

Alcohols with high carbon contents such as butanol can be synthesized from syngas through catalytic reaction that employs modified catalysts used in Fischer-Tropsch or methanol synthesis.

Feedstock for Alcohol Fuels

Alcohols fuels can be made from all available organic materials. Natural gas, coal, biomass, and organic wastes are good sources. Alcohol fuels have been synthesized from corn and sugar cane as major raw materials, but focal issues nowadays are synthesis and production of alcohol fuels from non-food crops and agricultural residues. Non-food lignocellulosic biomass includes energy crops, cellulosic residues, and wastes.

Grain-based ethanol as a first generation has been tried to change to the second-generation cellulosic ethanol and other advanced cellulosic biofuels. Cellulosic ethanol has identified as a key biochemical route of converting biomass to fuels after the 2000s. Algae-based third-generation feedstock for alcohol fuels emerged as a candidate that can provide a vast raw material for future alcohol fuel industry. Figure below illustrates the generations of raw feedstock for the alcohol fuel production and also shows the most apparent material that is being utilized in different countries.

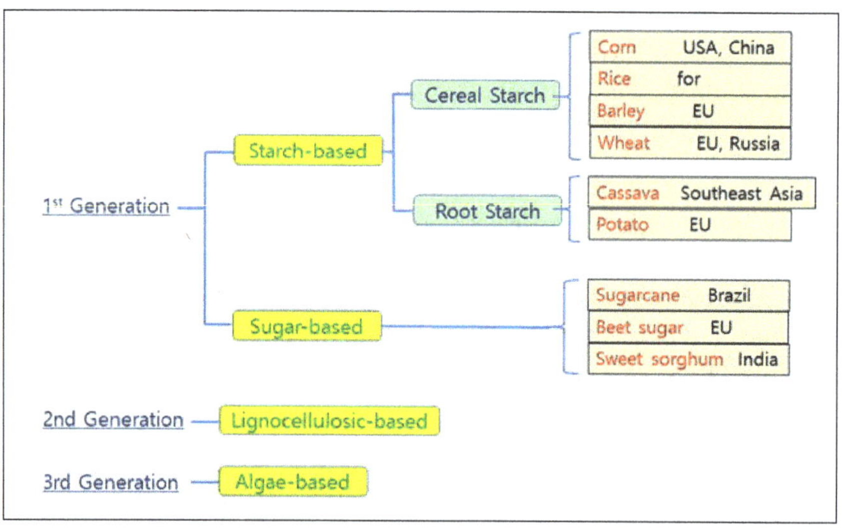

Fig: Key raw materials for bio-ethanol production in different countries.

Definitely there exists a clear difference between developing countries and developed countries in the priority choice, but basic understanding should be identical: use the locally available, underutilized feedstock, and choose the feedstock that tipping fee is available to treat the feedstock like municipal/industrial wastes. However, when wastes are involved as feedstock, it should be noted that not-in-my-backyard (NIMBY) problem occurs as a norm in almost every countries nowadays.

Utilization of Alcohol Fuels

About 66% of worldwide ethanol products are used for transportation purpose and

21% goes to industrial use. Most widely used area is gasohol in that alcohols are mixed to replace a portion of gasoline. In the near future, alcohol-using fuel cells and alcohol-mixed jet fuels are promising area of application.

Figure below shows the ethanol consumption trend for mixing to gasoline in the United States during the period of 1980–2020. During the years 2005–2010, ethanol use has drastically increased and remains as ca. 10% of the total gasoline consumed amount.

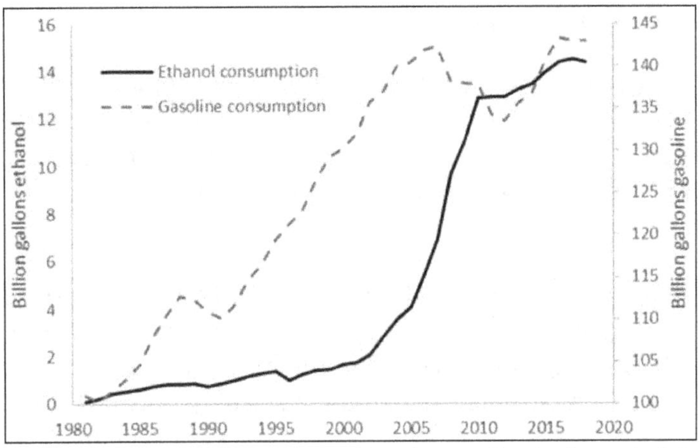

Fig: Trend of alcohol additive consumption for gasoline in the United States.

Fuel for Automobiles

The most abundant application of alcohol fuels is related to internal combustion engine of automobiles. Mixing alcohol fuels into gasoline has also a purpose of reducing pollutants by oxygenating the fuel. Since methanol is less expensive to produce than ethanol, although methanol is generally more toxic and has lower energy density than ethanol, it has been used with ethanol as automobile fuels. Compared to gasoline, methanol and ethanol have characteristics of burning at lower temperatures and lower volatility, which results in difficulty in starting the automobile engine in cold weather.

Current alcohol mixing status of ethanol-based fuel utilization in different countries is tabulated in table below.

Table: Ethanol mixing program to gasoline in different countries.

Country	Ethanol program
Brazil	Mandatory of bio-ethanol proportion as mixture of 24 ± 2%
United States	10% target for bio-ethanol proportion among primary energy sources (2010)
EU	Biofuel proportion increase to 2% in 2005, to 5.75% in 2010
Canada	Mandatory of bio-ethanol proportion in fuel set at 10%
China	Mandatory mixing of bio-ethanol at regional government level
India	Current mandatory 5% mixing of bio-ethanol and to be increased to 20%

Columbia	10% bio-ethanol to be mandatory at metropolitan area
Thailand	10% mandatory mixing of bio-ethanol to be enforced within Bangkok area
Argentina	5% bio-ethanol to be mandatory

Currently, flexible fuel vehicle (FFV) with dual fuel supply system for ethanol and gasoline is commercialized and widely distributed. In the case of methanol blending to gasoline, it is limitedly used in China from the 2000s. In China, M15 (15% methanol/85% gasoline) is the most familiar type.

Low-molecular weight alcohols such as ethanol have replaced conventional octane boosting additives like MTBE in automobile fuels. Alcohols that are added to gasoline make the mixed fuel to combust more completely by acting of higher oxygen content by alcohols and provide the ensuing effects of higher combustion efficiency and lower air pollution emissions.

In the United States, bio-ethanol is mandatorily mixed with transportation fuels. It has been reported that the bio-ethanol policy reduced crude oil reliance to 25% from 60%, and simultaneously creating 400,000 jobs, reducing 43% in greenhouse gas generation and cost-saving effect of $1.5/gallon-gasoline to the consumer.

Small amounts of methanol and higher alcohols are also allowed to be blended into gasoline within EN228 limits. E85 is used in FFVs in certain areas within the EU (such as Sweden, France, and Germany). There was a trial to use a near-neat fuel as M85 which contains 85% methanol/15% gasoline.

Racing cars used methanol for a long time, mainly by not producing black smoke which otherwise will block the view of ensuing other racing cars. Other than this application to racing cars, methanol fuel has not applied widespread other than some experience in China, methanol programs in California during the 1980–1990s, and a trial in Sweden as a marine fuel.

More than 98% of US gasoline contains typically 10% ethanol as E10 (10% ethanol/90% gasoline). Flexible fuel automobiles that can use E85 (85% ethanol/15% gasoline) exist in the United States and Brazil.

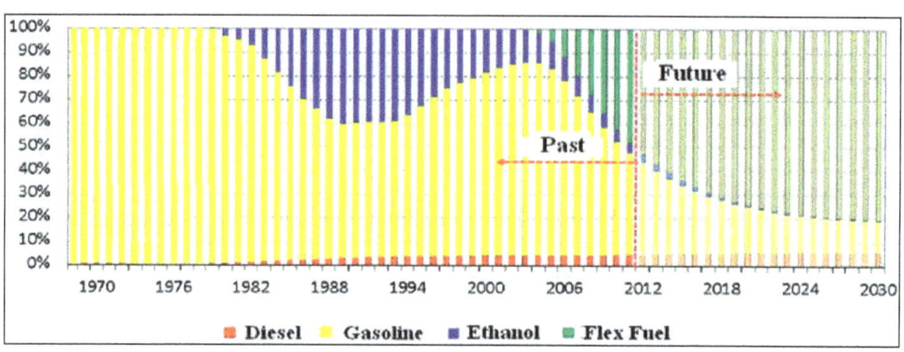

Fig: Changing trend of automobile fuel in Brazil.

In Brazil, 95% of automobiles are using fuel-flex engine system. Around 70% of automobiles in Brazil are able to run on ethanol, and the Brazil's demand for ethanol is estimated to increase by around 70% by 2030.

As for bio-butanol, the commercial scale production facility has not been constructed in sufficient numbers. In the United States, bio-butanol can be mixed up to 12.5%, and the 16% mixture is reported to be equivalently effective to existing E10.

Ethers such as dimethyl ether (DME) contain oxygen in chemical structure which acts as an oxidant in minimizing soot formation. Other exhaust emissions such as unburned hydrocarbons, NOx, and particulate matter are also reduced. DME is an ultra-clean fuel that has similar properties to LPG.

Fuel Cells

Alcohol fuel cell is an energy conversion device to generate electricity via electrochemical reactions on the catalytically active electrodes without direct combustion of alcohol fuel. Direct alcohol fuel cell (DAFC) is named for its direct supply of alcohol to fuel electrode and is called with specific terminologies as direct methanol fuel cell or direct ethanol fuel cell depending on the alcohol fuel source.

The mechanism of electricity generation is based on the oxidation of methanol fuel at the anode (fuel electrode) and conduction of electron(s) to the cathode (air or oxygen electrode) via external conducting circuit and simultaneous electrolytic conduction of proton (H^+) via polymer electrolyte to the cathode.

DAFC can provide portable energy source to electronic devices such as cellular phones and notebook computers. DAFC that uses alcohol can have several advantages in terms of storage, transportation, safety, etc. over fuel cell systems like proton exchange membrane fuel cell (PEMFC) that use hydrogen.

Jet Fuel

Alternative jet fuel typically contains a complex mixture of primarily n/iso-paraffins, cycloparaffins, and alkylbenzenes with a carbon number range of 9–15. Carbon tax accelerates the development of jet fuel from renewable resources. Lowering emissions of particles and greenhouse gases during the flight are the fundamental reason of trying alcohol fuels as a jet fuel option.

Environmental Aspects

Global recognition regarding the significant long-term impact due to climate change provides a key foundation for utilizing alcohol fuels, which means that alcohol fuels should be able to accommodate chances in reducing climate change gases(CO_2, methane, N_2O, etc.). Bio-ethanol is highly effective in reducing greenhouse gas evolution. Corn-based bio-ethanol is reported to generate 43% less greenhouse gases compared to pure gasoline.

From purely theoretical point of view, ethanol can be finally produced from the biomass that is made based on the CO_2 absorbed by plants during photosynthesis, and thus it can be called carbon neutral. Unlike hydrocarbons which evolve voluminous amount of CO_2 from their internal carbon atoms during combustion, ethanol can be regarded as carbon neutral without generating as much CO_2 from internal carbon atoms. In practice, however, significant amount of greenhouse gas evolution is directly and indirectly caused by cultivation of biomass crops and synthesis of alcohol fuels.

The range of CO_2 reduction potential is large when alcohol fuels are used. Values range between 0.5 kg CO_2-equivalent/liter of ethanol for ethanol produced from wheat and up to 2.24 kg CO_2-equivalent/liter of ethanol for ethanol manufactured from sugar cane.

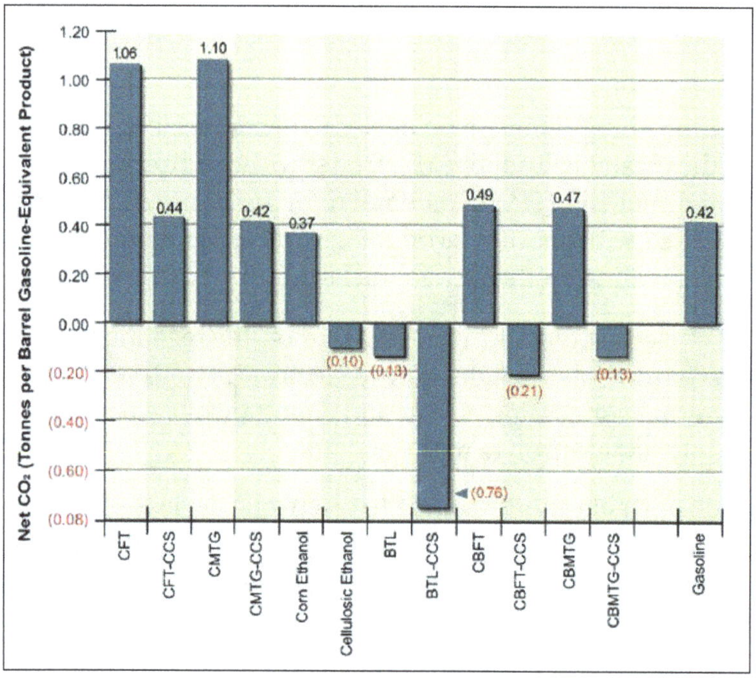

Estimated carbon dioxide (CO_2) emissions over the life cycle of alternative fuels. BTL, biomass to liquid; CBFT, coal and biomass to liquid, Fisher-Tropsch; CBMTG, coal and biomass to liquid, methanol to gasoline; CCS, carbon capture and storage; CFT, coal to liquid, Fisher-Tropsh; CMTG, coal to liquid, methanol to gasoline.

According to the result shown in figure above, among alternative liquid fuels, only cellulosic ethanol, biomass to liquid (BTL), and CCS-involved processes (BTL-CCS, CBFT-CCS, CBMTG-CCS) exhibit the CO_2-negative performance in life cycle analysis (LCA) perspective. Carbon capture and storage (CCS) process is not fully economically feasible and technically proven till now; moreover, considering public objection on CCS, connecting the process to CCS is not practical for the time being. In CO_2 reduction aspect, cellulosic ethanol is the most reasonable choice as a renewable alternative fuel.

Adding ethanol to gasoline fuel of automobile, oxygen contents of fuel mixture increases and yielding the effect of reducing pollutants evolution. As alcohol fuels are inherently sulfur-free, it suits for cleaner environment. Besides, since ethanol is produced by fermentation with crops that contains starch, its purity is high, and no hazardous combustion by-products such as SO_2 or metal oxides are generated during the combustion when compared to the petroleum-based fuel. But, high solubility into water by short carbon chain alcohols such as ethanol and methanol can cause underground water pollution, although short carbon chain alcohols are well degradable in few days under normal circumstances. This problem can be minimized with the use of higher carbon chain alcohols like bio-butanol.

Forest clearing and chemical fertilizer are involved to grow corn and other grains for the first-generation bio-ethanol, which eventually ends up with CO_2 production and countervailing the CO_2 reduction amount by bio-ethanol use, sometimes more than the reduced amount. In this regard, the second-generation lignin-based or third-generation algae-based raw material is a better candidate for bio-alcohols.

It is especially noteworthy that the definition of (environment friendly) bio-energy is rather more stringently defined in the EU and United States: more than 35% reduction of greenhouse gas is required to qualify compared to fossil fuels of the same calorific value.

The issue of required water amount asks the approach of water-energy nexus in that technology development will follow for the process of better environment-friendliness and sustainability. As an example, recent water shortage encountered in Chennai, India, might be attributable to global warming, and the water quantity consumed for the production of alcohol fuels is emerging as an important issue. Chennai region went without rain for 200 days in 2018.

The process consuming the largest amount of water is the cultivation of biomass crops. Among the production processes for alcohol fuels, refinery step consumes the largest amount of water. The water quantity consumed for US corn-based ethanol production is approximately equivalent to the water requirement that can sustain 5000 people for 1 year.

Moreover, the refinery process that is going to be extended for the second-generation cellulosic ethanol is expected to consume 2.9 times more water needed for corn-based ethanol refinery process. At present, cellulosic ethanol production process consumes about 9.8 L/L-ethanol, which is unduly high.

Aviation Biofuel

Aviation accounts for around 15% of global oil demand growth up to 2030 in the IEA's New Policies Scenario, a similar amount to the growth from passenger vehicles. Such a rise means that aviation will account for 3.5% of global energy related CO_2 emissions by 2030, up from just over 2.5% today, despite ongoing improvements in aviation efficiency.

This expansion underscores the need for the aviation industry to tackle its carbon emissions. For now, liquid hydrocarbon fuels like jet fuel remain the only means of powering commercial air travel. Therefore, along with a sustained improvement in energy efficiency, Sustainable Aviation Fuel (SAF) such as aviation biofuels are key to reducing aviation's carbon emissions.

The aviation industry has committed to reducing carbon emissions by 50% from their 2005 level by 2050. Blending lower carbon SAF with fossil jet fuel will be essential to meeting this goal. This is reflected in the IEA's Sustainable Development Scenario (SDS), which anticipates biofuels reaching around 10% of aviation fuel demand by 2030, and close to 20% by 2040.

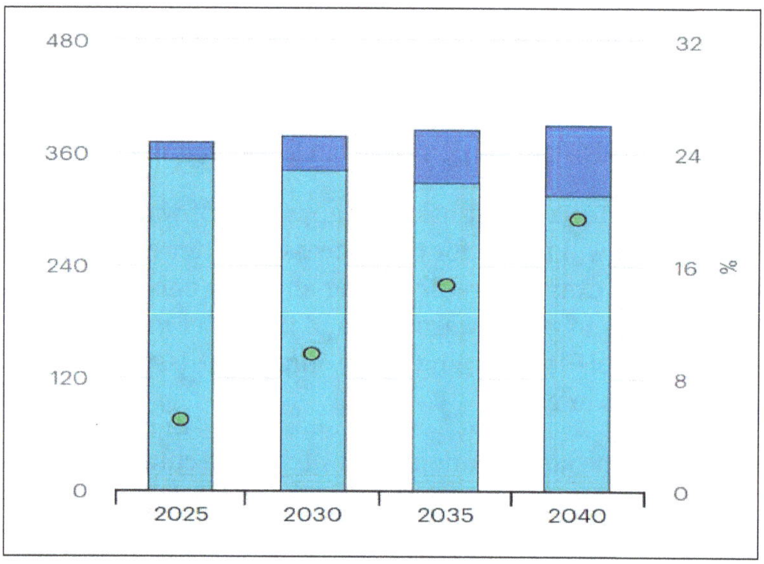

Fig: Aviation fuel consumption in the Sustainable Development Scenario, 2025-2040.

The first flight using blended biofuel took place in 2008. Since then, more than 150,000 flights have used biofuels. Only five airports have regular biofuel distribution today (Bergen, Brisbane, Los Angeles, Oslo and Stockholm), with others offering occasional supply. But the centralised nature of aviation fuelling, where less than 5% of all airports handle 90% of international flights, means SAF availability at a small number of airports could cover a large share of demand.

Another indication of aviation's commitment to growing SAF use is the agreement of long-term off-take agreements between airlines and biofuel producers. These now cumulatively cover around 6 billion litres of fuel. Meeting this demand will require further production facilities, and some airlines have directly invested in aviation biofuel refinery projects.

Still, aviation biofuel production of about 15 million litres in 2018 accounted for less than 0.1% of total aviation fuel consumption. This means that significantly faster market development is needed to deliver the levels of SAF production required by the aviation industry and keep on track with the requirements of the SDS.

Currently, five aviation biofuel production pathways are approved for blending with fossil jet kerosene. However, only one – hydroprocessed esters and fatty acids synthetic paraffinic kerosene (HEFA-SPK) fuel – is currently technically mature and commercialised. Therefore, HEFASPK is anticipated to be the principal aviation biofuel used over the short to medium term.

Meeting 2% of annual jet fuel demand from international aviation with SAF could deliver the necessary cost reduction for a self-sustaining aviation biofuel market thereafter. Meeting such a level of demand requires increased HEFA-SPK production capacity. If met entirely by new facilities, approximately 20 refineries would be required. This could entail investment in the region of $10 billion. Although significant, this is relatively small compared to fossil fuel refinery investment of $60 billion in 2017 alone.

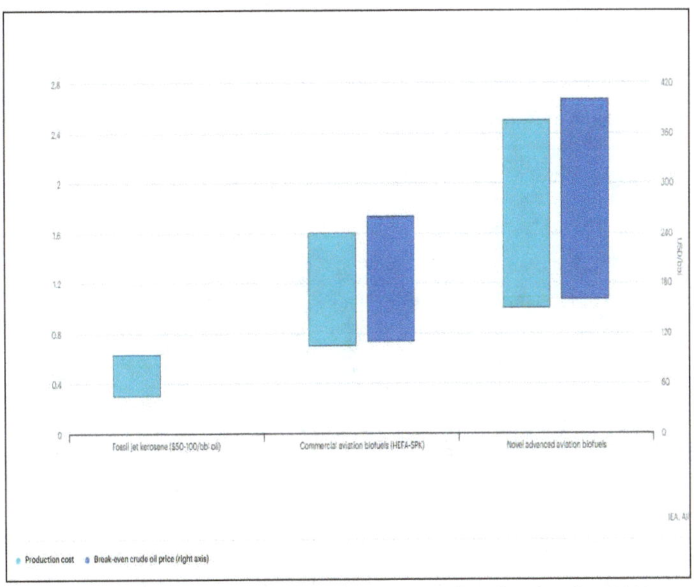

Fig: Production cost and break-even crude oil price for SAFs compared with fossil jet kerosene, 2019.

Ongoing research and development is needed to support the commercialisation of novel advanced aviation biofuels which can unlock the potential to use agricultural residues and municipal solid wastes. These feedstocks are more abundant and generally cost less than the waste oils and animal fats commonly used by HEFA-SPK, and can therefore facilitate greater SAF production. Furthermore, synthetic fuels produced from renewable electricity, CO_2 and water via Power-to-Liquid processes may offer an alternative fuel source for aviation in the long term.

SAF are currently more expensive than jet fuel, and this cost premium is a key barrier to their wider use. Fuel cost is the single largest overhead expense for airlines, accounting for 22% of direct costs on average, and covering a significant cost premium to utilize aviation biofuels is challenging.

The competitiveness of SAF depends on its production cost relative to that of fossil jet kerosene (which varies with crude oil price). For all biofuels obtaining an economic feedstock supply is fundamental to achieve competitiveness, as feedstocks are the major determinant of production costs. For HEFA-SPK economies of scale could be realized by refineries designed for continuous production.

In the long term, airlines may include SAF consumption cost premiums within ticket costs. At current prices and today's fleet average energy efficiency, the additional cost per passenger for a 15% blend of HEFA may not be high in comparison with other elements that influence ticket prices, such as seating class, the time of ticket purchase and taxation. However, due to the competiveness of the aviation industry customer price sensitivity is a core consideration for airlines.

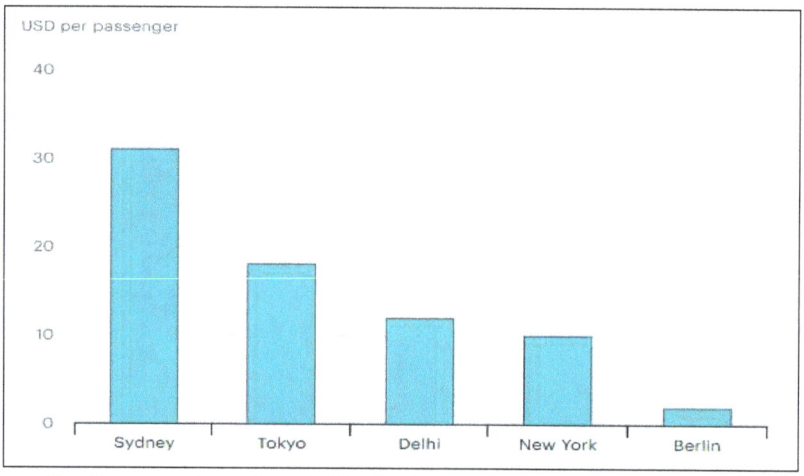

Fig: Cost premium of a 15% blend of commercial aviation biofuels per apssagener from London to selected cities, 2019.

Impressive progress has been made in the utilisation of SAF since the first biofuel flight ten years ago. However, to fulfill aviation biofuels' potential to reduce the climate impact of growing air transport demand, further technological development and improved economics are needed.

There is a key role for policy frameworks at this crucial early phase of SAF industry development. Without a supportive policy landscape, the aviation industry is unlikely to scale up biofuel consumption to levels where costs fall and SAF become self-sustaining.

Technologies for Bio Jet Fuel

Representative production technologies for bio-jet fuel include alcohol-to-jet (ATJ), oil-to-jet (OTJ), gas-to-jet (GTJ), and sugar-to-jet (STJ) process. OTJ process produces bio-jet fuel from animal or plant tallow such as waste vegetable oil, beef tallow, and microalgae. More specifically, hydrotreated esters and fatty acid (HEFA) technology, a kind of OTJ process, encompasses hydrotreated renewable jet (HRJ) process among HEFA technologies, catalytic hydro-thermolysis (CH), and rapid thermal decomposition process (HDCJ). STJ process involves catalytic upgrading and conversion of glucose- or starch-based raw material to hydrocarbons or biological conversion to bio-jet fuel via direct sugar to hydrocarbons (DSHC) and catalytic upgrading. ATJ process involves production of bio-jet fuel via hydrolysis of wooden fiber biomass or glucose into intermediate alcohols (methanol, ethanol, butanol, and fatty acid alcohols) and their dehydration and oligomerization. It is divided into ethanol-to-jet or butanol-to-jet technologies, depending on alcohol involved. GTJ process involves biogas, natural gas, or syngas from wood fiber biomass to bio-jet fuel via bio-chemical or thermos-chemical routes such as gas fermentation and Fischer-Tropsch processes. Table below shows the production technologies for bio-jet fuel.

Technologies	Production processes
Alcohol to jet	Ethanol to jet
	N-butanol to jet
	Iso-butanol to jet
	Methanol to jet
Oil to jet	Hydro-processed renewable jet
	Catalytic hydro-thermolysis
	Hydrotreated depolymerized cellulosic jet
Gas to jet	Fischer-Tropsch synthesis
	Gas fermentation
Sugar to jet	Direct sugar to hydrocarbons
	Catalytic upgrading

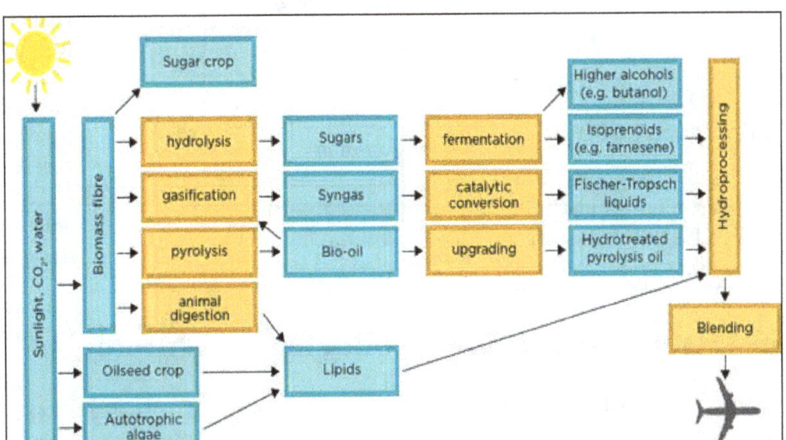

Fig: Production process-wise raw material and technology overview for bio-jet fuel.

Among many classification methods, bio-jet fuel is divided via production pathways: fermentation, deoxidation, or thermal decomposition. As of 2016, ASTM 7566 dictates five production processes (Fischer-Tropsch Synthetic Kerosene with Aromatics (FT-SPK), HEFA, Synthesized Iso-Paraffins (SIP), ATJ) as means to produce commercially viable bio-jet fuels. It simultaneously regulates product quality criteria as per 100% assay as well as mixing proportion in existing petroleum-based aviation fuel. Many other production processes are also used to produce bio-jet fuel.

Table: Production process-wise classification of bio-jet fuel production process.

Production process	Developer/manufacturer	Raw materials	Aromatic content	ASTM review stage and max. Mixing proportions
FT-SPK	Sasol, Shell, Syntroleum	Coal, natural gas, biomass	Low	(2009)-50% Approved
HEFA	Honeywell UOP, Neste Oil, Dynamic Fuels, EERC	Vegetable oil, animal fat, recycled vegetable oil	Low	(2011)-50% Approved
SIP	Amyris, Total	Sugar	Low	(2014)-10% Approved
ATJ-SPK	Gevo, Cobalt, Honeywell UOP, LanzaTech, Swedish Biofuels, Byogy	Starch, sugar, cellulose-based biomass	Low	(2016)-30% Approved
FT-SKA	Sasol	Coal, natural gas, biomass	High	Under review by committee
HDO-SK	Virent	Starch, sugar, cellulose-based biomass	Low	Investigation report submitted
HDO-SAK	Virent	Starch, sugar, cellulose-based biomass	High	Investigation report under review
HDCJ	Honeywell UOP, Licella, KiOR	Cellulose-based biomass	High	Supplement to investigation report received
CH	Chevron Lummus Global, Applied Research Associates, Blue Sun Energy	Vegetable oil, animal fat, recycled vegetable oil	Low	Investigation report under review

Bio-Alcohol-Based Bio-Jet Fuel Production Technologies

Figure below shows current worldwide production and consumption trend of bio-jet fuel. Bio-ethanol is widely commercialized as sustainable source of energy for use in transportation with worldwide production of 104 million m^3 and 80% of its utilization as transportation fuel. The USA and Brazil accounted for 51.8 and 2.77 million m^3 productions, respectively. Worldwide bio-jet fuel amounted to 30 billion m^3. On the other hand, Korean domestic petroleum-based aviation fuel products totaled 13% (20.66 million m^3) in 2013, which is similar to gasoline products (13.5%) and 44% of light oil products (29.7%).

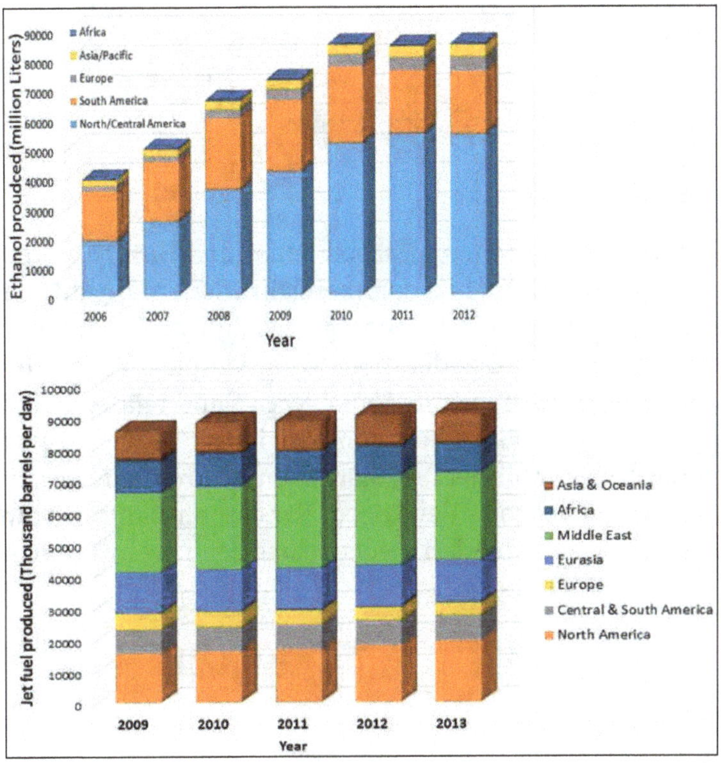

Fig: Worldwide bio-ethanol production and consumption of aviation fuel.

Possible raw material for ATJ process includes methanol, ethanol, and butanol. Such alcohol-based raw material is converted to bio-jet fuel via polymerization and upgrading technology. Among these alcohols, bio-ethanol utilization is promising in view of its current production and consumption and worldwide use. At present, bio-ethanol is mixed to maximum 10~15% with gasoline. Although potential market of ethanol for mixing with gasoline seems limited for expansion, conversion to bio-jet fuel via bio-ethanol upgrading shows possibility of replacing existing petroleum-based aviation fuel.

For conversion of bio-ethanol to bio-jet fuel, physicochemical properties of bio-ethanol should be compatible with petroleum-based aviation fuel. The USA is utilizing advanced ATJ technology to make physicochemical properties of bio-ethanol compatible with those of existing petroleum-based fuel. More specifically, 99.5~99.9% of anhydrous ethanol is mixed with existing fuel or converted to bio-jet fuel. High purity ethanol is used as raw material in the process for upgrading physicochemical properties of bio-jet fuel. Such ATJ process is based on bio-ethanol for production of bio-jet fuel, and oxygen contents of bio-ethanol is removed by dehydration, polymerization for access of carbon atoms from existing petroleum-based aviation fuel, and hydrogenation reaction for optimization of physicochemical properties. Figure below shows technical overview of ATJ process for production of bio-jet fuel from bio-ethanol.

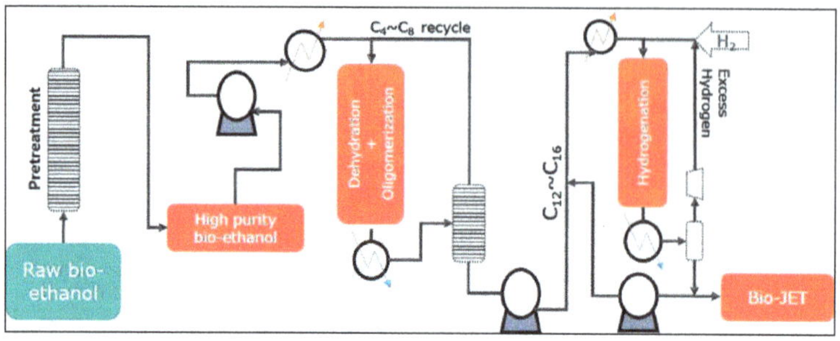

The most efficient method of reducing carbon emission is low carbon bio-jet fuel, relevant technology to produce it and its commercialization. Many international airlines initiated small-scale projects, but so far economic viability has not been demonstrated, and possible remedy is under consideration. To accomplish such economic viability, international standards for carbon emission objective and related policy on the part of airlines have to be established as well as monetary incentive for bio-jet fuel utilization. To initiate economic drive for bio-jet fuel market, mass production-capable technology for bio-jet fuel production has to be developed. For carbon-neutral growth by 2050, international carbon emission reduction objective has been set by the ICAO with respect to greenhouse gas emission of 2005.

Syngas

The synthesis gas is defined as a gas with H_2 and CO as the main components of fuel. Row syngas contains mainly significant amounts of CO_2 and H_2O as well. Since syngas is usually used at higher pressures for synthesizing chemicals and fuels, the N_2 contents must usually be minimized in syngas. Bio-syngas, however, are biomass produced, chemically identical to syngas. This definition is in accordance with the SYNBIOS-conference definition.

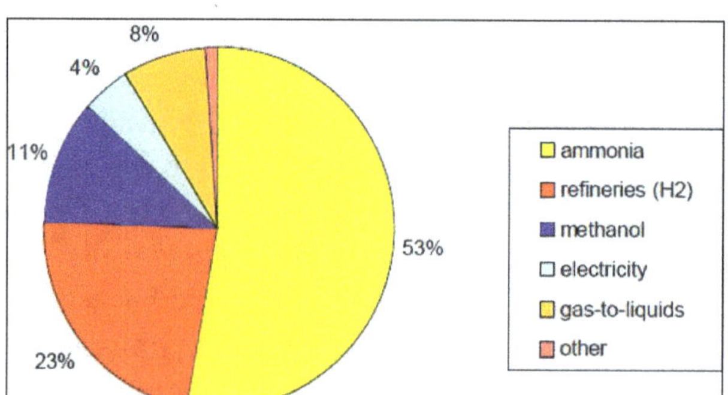

Fig: Present world syngas market, totally ~6 EJ/y.

Syngas differ chemically from gasses normally generated by gasification processes at low temperatures, including fluidized bed reactors. To be clear, the gas produced by such reactors is referred to in this book as "product gas." Product gas is defined as a fuel gas with H_2 and CO as well as with substantial amounts of hydrocarbons, such as methane. Product gas contains CO_2 and H_2O, and often N_2, also inevitably.

Throughout the chemical industry, syngas is a substantial intermediate product. Each year, around 6 EJ of syngas are manufactured globally, which is almost 2% of the world's current primary energy consumption. The ammonia industry dominates the global market for syngas (mainly from fossil fuels like coal, natural gas, and oil/residues). The production of hydrogen for use in refineries, for instance, the processing of hydrogen, and methanol are other major applications. The current market distribution of syngas is shown in above figure.

Major Properties of Syngas

Different characteristics of syngas can affect the process of combustion in internal combustion (IC) motors. The flammability limit of the syngas is one of the most important properties in IC engine safety and fuel. Also, the laminar flame velocity (burning velocity) is an essential parameter to investigate the operation of the combustion chamber and its emission performance.

Syngas Flammability Limits

The limit of flammability is usually used as an index for the flammability of the gas. This describes the range of the fuel concentrations in the fuel/air mixture at certain temperature and pressure, which allow the ignition of the flame to propagate and sustain the flammability limits are known in line with generally accepted usages as those fuel-air areas where flame propagation can take place and where fire cannot propagate. The fuel, the spread direction, the size and the form of the combustion chamber, the temperature, and the pressure are primarily affected. And for the fuel-air blend, there are two distinct flammability limits, namely the smallest fuel boundary the flames can propagate is called the lower flammability boundary (LFL), while the richest one is called the upper flammability boundary (UFL). The fact that H_2 and CO are the principal flame-retardant components of syngas inherits the characteristics of these gasses. The presence of inert gasses such as nitrogen and carbon dioxide in the gas mixtures reduces the flammability limit.

Laminar Flame Velocity

The laminar flash speed is the speed at which the flame propagates in the direction of expansion wave surfaces under a laminar flow condition via quiet unbranded fuel-oxidant mixes. Because LFV is highly sensitive to combustion chamber operations and emission performance, it is very important for the investigation of combustion chamber operations. The composition of the fuel, mixture equivalence ratio, temperature, and pressure affects it.

Syngas Composition and its Calorific Value

The composition of the manufacturer's gas depends on feedstock, particulate size, gas flow rate and feedstock flow, chemical reactor configurations, operating conditions or process of gasification, gasificator and catalyst, and gas residence time. But the temperature of the reactor, which in turn is affected by the ER value, mainly influences it. Furthermore, CO, H_2, and CH_4 concentrations in producer gas are also controlled by chemical reactions in the process of gasification.

There is, therefore, a considerable influence on the calorific value of the producer gas on the type of oxidizing agent used for gasification. As ER increases and then the concentrations of these useful components decreases because of the intensification of combustion at higher ER values, the concentrations of CO and H_2 reach the maximum value. As ER increases, the concentrations of CO_2 and N_2 in the producer gas are also increased. Air as an oxidant produces syngas with relatively high levels of nitrogen and thus a lower heat value, which does not normally exceed 6 MJ/Nm^3.

The producer's gas is classified as fuel gas of low quality. The typical biomass gasification composition of an air-borne downdraft reactor with the oxidizer is as follows: 15–20% of H_2, 15–20% of CO, 0.5–2% of CH_4, 10–15% of CO_2, and the rest of the component of N_2, O_2, and CXHY. If the concentration of fuel components is considerately increased and the gas is called a medium heat value, up to 16 MJ/Nm^3, where oxygen or water steam or the mixture of both are used.

Production of Synthesis Gas

Synthesis gas or syngas is called carbon monoxide (CO)-hydrogen (H_2)-containing gas mixture. Carbon dioxide (CO_2) and other components such as water (H_2O) may also be present in syngas.

The chemical synthesis can be used as a building block in all products normally produced from crude oil or natural gas. Petrol and diesel are fuels with definitions, which are not based on the chemistry but on their physical qualities such as boiling and flashing.

The octane rating of gasoline in an internal combustion (IC) engine is empirical and based on its actual performance. This means that at least low-grade blending is possible as long as synthetic fuel matches the characteristics of petrol- or diesel-based crude oil.

Production of Fuels and Chemicals from Gasification of Biomass/Coal or Reforming of Natural Gas

The feedstock must be gasified when beginning with a solid feedstock, such as biomass or coal. It may be necessary to ground or pulverize the feedstock before gasification (usually carbon). The particle fineness depends on the gasification type. For most biomass gasification plants, drying is required as the next step. The drying is integrated into a gasifier reactor vessel by certain gasifiers.

Following gasification, the product is a gas, known as producer gas, filled with impurities that must be removed. Prior to synthesis, the producer of gas usually also needs the ratio from H_2 to CO. H_2:CO both.

Gasification

All hydrocarbon feed resources like coal, heavy oils, or combustible biomass can be gasified as synthesis gas. Several reactions occur in the gasificator, but the total reaction can be summarized by two equations given below.

$$\text{Biomass} + O_2 \rightarrow CO + H_2 + CO_2 + H_2O + CH_4$$

By-products: tar, char, ashes.

Reaction conditions:

- High temperature (800–1000°C) and low pressure (1–20 bar).
- H_2/CO ratio within 0.5 and 1.8, depending on the technologies.

The reaction also can be expressed as:

$$\text{CaHbOcNd} + O_2 / H_2O / N_2 \rightarrow CO + H_2 + \text{CxHyOz} + CO_2 + H_2O + NH_3 + N_2$$

The solid carbon is partially oxidized with oxygen (O_2), air, steam (H_2O), or a combination of all gasification agents. CxHyOz is mostly made of methane with a few low percent of hydrocarbon, including ethane and ethylene. For most gasifiers, gas may also contain heavier hydrocarbons such as benzene, toluene, and naphthalenes, depending on the feedstock and the operational parameters. Hydrocarbons, which are heavier than benzene, are often known as tars.

Table: Composition of producer gas.

Composition	Updraft Gasifier (% by volume)	Downdraft Gasifier (% by volume)
Carbon monoxide	24	21
Hydrogen	11	17
Methane	3	2
Hydrocarbon	0.2	0.3
Nitrogen	53	48
Water vapor	3	4

This is an important factor in determining the technical mechanism and the economic feasibility of the gasification system. Efficiencies in gasification are based on the biomass type used, its particle size, its ER value, and the reactor design.

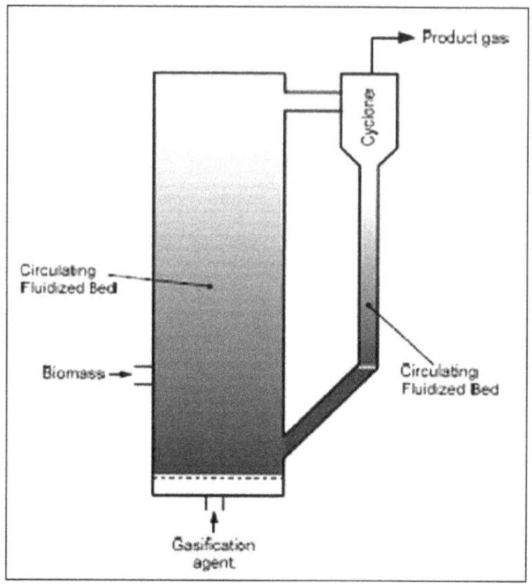

Fig: The gasificator.

The gasification efficiency is usually determined on the lower heating value basis. The efficiency is calculated as the ratio of the total energy in the producer gas (sensible and chemical) and the chemical energy in the feedstock (the heating value). Depending on type and design of the gasifier as well as on the characteristics of the fuel, mechanical gasifier efficiency may vary between 60 and 75%. A useful definition of the gasification efficiency (%) used for engine applications is as follows:

$$\eta_m = \frac{H_g \times Q_g}{H_s \times M_s} \times 100$$

Where, η_m is the gasification efficiency (%) (mechanical), Hg is the heating value of the gas (kJ/m³), Qg is the volume flow of gas (m³/s), H_s is the lower heating value of gasifier fuel (kJ/kg), and M_s is the gasifier solid fuel consumption (kg/s).

The Yield of Syngas

The syngas yield is measured by the mass of the produced in cubic meters per the mass of the feedstock provided to the system. The yield is directly commensurate with the difference in ER and with the gas residence time in the reduction area. The biomass ash content also has a considerable impact and limits the yield of the gas producer.

Cleaning and Cooling of Producer Gas

The combustible gas can be used as a feedstock for the production of chemicals like methanol or in internal combustion engines for direct heat uses.

Biodiesel

Biodiesel is an alternative fuel similar to conventional or 'fossil' diesel. Biodiesel can be produced from straight vegetable oil, animal oil/fats, tallow and waste cooking oil. The process used to convert these oils to Biodiesel is called transesterification. The largest possible source of suitable oil comes from oil crops such as rapeseed, palm or soybean. In the UK rapeseed represents the greatest potential for biodiesel production. Most biodiesel produced at present is produced from waste vegetable oil sourced from restaurants, chip shops, industrial food producers such as Birdseye etc. Though oil straight from the agricultural industry represents the greatest potential source it is not being produced commercially simply because the raw oil is too expensive. After the cost of converting it to biodiesel has been added on it is simply too expensive to compete with fossil diesel. Waste vegetable oil can often be sourced for free or sourced already treated for a small price. (The waste oil must be treated before conversion to biodiesel to remove impurities). The result is Biodiesel produced from waste vegetable oil can compete with fossil diesel.

Benefits of Biodiesel

Biodiesel has many environmentally beneficial properties. The main benefit of biodiesel is that it can be described as 'carbon neutral'. This means that the fuel produces no net output of carbon in the form of carbon dioxide (CO_2). This effect occurs because when the oil crop grows it absorbs the same amount of CO_2 as is released when the fuel is combusted. In fact this is not completely accurate as CO_2 is released during the production of the fertilizer required to fertilize the fields in which the oil crops are grown. Fertilizer production is not the only source of pollution associated with the production of biodiesel, other sources include the esterification process, the solvent extraction of the oil, refining, drying and transporting. All these processes require an energy input either in the form of electricity or from a fuel, both of which will generally result in the release of greenhouse gases. To properly assess the impact of all these sources requires use of a technique called life cycle analysis. Biodiesel is rapidly biodegradable and completely non-toxic, meaning spillages represent far less of a risk than fossil diesel spillages. Biodiesel has a higher flash point than fossil diesel and so is safer in the event of a crash.

Biodiesel Production

Biodiesel can be produced from straight vegetable oil, animal oil/fats, tallow and waste oils. There are three basic routes to biodiesel production from oils and fats:

- Base catalyzed transesterification of the oil.
- Direct acid catalyzed transesterification of the oil.
- Conversion of the oil to its fatty acids and then to biodiesel.

Almost all biodiesel is produced using base catalyzed transesterification as it is the most economical process requiring only low temperatures and pressures and producing a 98% conversion yield.

The Transesterification process is the reaction of a triglyceride (fat/oil) with an alcohol to form esters and glycerol. A triglyceride has a glycerine molecule as its base with three long chain fatty acids attached. The characteristics of the fat are determined by the nature of the fatty acids attached to the glycerine. The nature of the fatty acids can in turn affect the characteristics of the biodiesel. During the esterification process, the triglyceride is reacted with alcohol in the presence of a catalyst, usually a strong alkaline like sodium hydroxide. The alcohol reacts with the fatty acids to form the mono-alkyl ester, or biodiesel and crude glycerol. In most production methanol or ethanol is the alcohol used (methanol produces methyl esters, ethanol produces ethyl esters) and is base catalyzed by either potassium or sodium hydroxide. Potassium hydroxide has been found to be more suitable for the ethyl ester biodiesel production; either base can be used for the methyl ester. A common product of the transesterification process is Rape Methyl Ester (RME) produced from raw rapeseed oil reacted with methanol.

The figure below shows the chemical process for methyl ester biodiesel. The reaction between the fat or oil and the alcohol is a reversible reaction and so the alcohol must be added in excess to drive the reaction towards the right and ensure complete conversion.

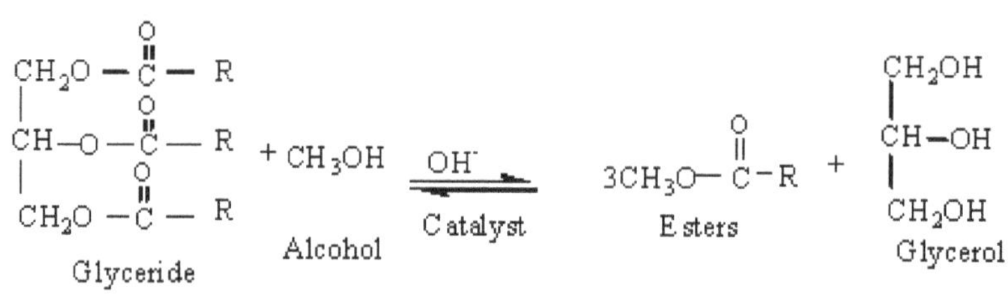

The products of the reaction are the biodiesel itself and glycerol.

A successful transesterification reaction is signified by the separation of the ester and glycerol layers after the reaction time. The heavier, co-product, glycerol settles out and may be sold as it is or it may be purified for use in other industries, e.g. the pharmaceutical, cosmetics etc.

Straight vegetable oil (SVO) can be used directly as a fossil diesel substitute however using this fuel can lead to some fairly serious engine problems. Due to its relatively high viscosity SVO leads to poor atomisation of the fuel, incomplete combustion, coking of the fuel injectors, ring carbonisation, and accumulation of fuel in the lubricating oil. The best method for solving these problems is the transesterification of the oil.

The engine combustion benefits of the transesterification of the oil are:

- Lowered viscosity.

- Complete removal of the glycerides.
- Lowered boiling point.
- Lowered flash point.
- Lowered pour point.

Production Process

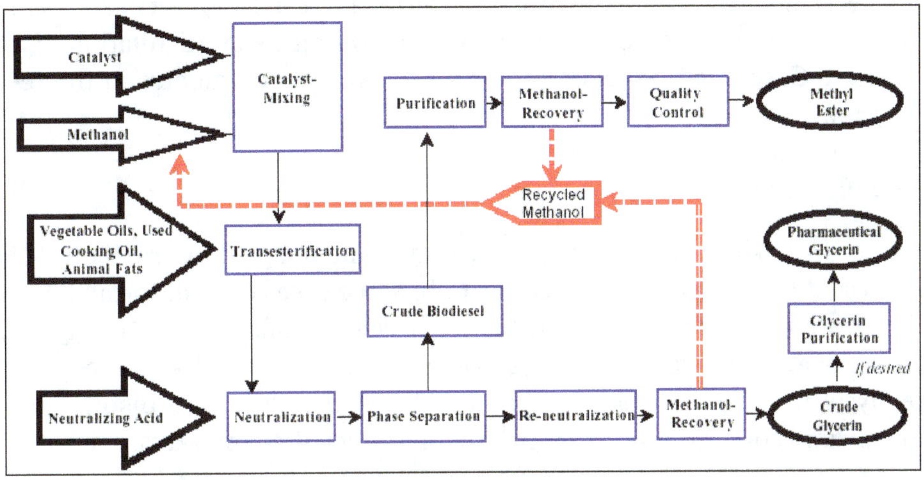

An example of a simple production flow chart is proved below with a brief explanation of each step.

Mixing of Alcohol and Catalyst

The catalyst is typically sodium hydroxide (caustic soda) or potassium hydroxide (potash). It is dissolved in the alcohol using a standard agitator or mixer. Reaction- The alcohol/catalyst mix is then charged into a closed reaction vessel and the oil or fat is added. The system from here on is totally closed to the atmosphere to prevent the loss of alcohol. The reaction mix is kept just above the boiling point of the alcohol (around 160 °F) to speed up the reaction and the reaction takes place. Recommended reaction time varies from 1 to 8 hours, and some systems recommend the reaction take place at room temperature. Excess alcohol is normally used to ensure total conversion of the fat or oil to its esters. Care must be taken to monitor the amount of water and free fatty acids in the incoming oil or fat. If the free fatty acid level or water level is too high it may cause problems with soap formation and the separation of the glycerin by-product downstream.

Separation

Once the reaction is complete, two major products exist: glycerin and biodiesel. Each has a substantial amount of the excess methanol that was used in the reaction. The reacted mixture is sometimes neutralized at this step if needed. The glycerin phase is much denser than biodiesel phase and the two can be gravity separated with glycerin simply drawn off the bottom of the settling vessel. In some cases, a centrifuge is used to separate the two materials faster.

Alcohol Removal

Once the glycerin and biodiesel phases have been separated, the excess alcohol in each phase is removed with a flash evaporation process or by distillation. In others systems, the alcohol is removed and the mixture neutralized before the glycerin and esters have been separated. In either case, the alcohol is recovered using distillation equipment and is re-used. Care must be taken to ensure no water accumulates in the recovered alcohol stream.

Glycerin Neutralization

The glycerin by-product contains unused catalyst and soaps that are neutralized with an acid and sent to storage as crude glycerin. In some cases the salt formed during this phase is recovered for use as fertilizer. In most cases the salt is left in the glycerin. Water and alcohol are removed to produce 80-88% pure glycerin that is ready to be sold as crude glycerin. In more sophisticated operations, the glycerin is distilled to 99% or higher purity and sold into the cosmetic and pharmaceutical markets.

Methyl Ester Wash

Once separated from the glycerin, the biodiesel is sometimes purified by washing gently with warm water to remove residual catalyst or soaps, dried, and sent to storage. In some processes this step is unnecessary. This is normally the end of the production process resulting in a clear amber-yellow liquid with a viscosity similar to petrodiesel. In some systems the biodiesel is distilled in an additional step to remove small amounts of color bodies to produce a colorless biodiesel.

References

- What-is-Biogas: homebiogas.com, Retrieved 08, March 2020
- Fact-sheet-biogas-converting-waste-to-energy: eesi.org, Retrieved 23, July 2020
- Alcohol-fuels-current-status-and-future-direction, alcohol-fuels-current-technologies-and-future-prospect: intechopen.com, Retrieved 19, February 2020
- Are-aviation-biofuels-ready-for-take-off: iea.org, Retrieved 05, May 2020
- Syngas-production-properties-and-its-importance, sustainable-alternative-syngas-fuel: intechopen.com, Retrieved 12, August 2020
- Woody-biomass-properties: extension.org, Retrieved 26, April 2020

3

Biofuels and their Sources

Biofuels can be derived from a large variety of sources. Some of the different types of biofuels on the basis of their sources are sugarcane biofuel, soybean biofuel, camelina biofuel, rapeseed biodiesel, cellulosic biofuel and animal fat biodiesel. This chapter discusses in detail these biofuels as well as their sources.

Sugarcane Biofuel

Sugarcane is any of 6 to 37 species (depending on taxonomic interpretation) of tall perennial grasses of the genus *Saccharum* (family Poaceae, tribe Andropogoneae). Sugarcane is native of warm temperate climate, common in the tropical regions as Brazil, India, Africa, and Asia pacific. The sugarcane plant morphology can be seen in figure below.

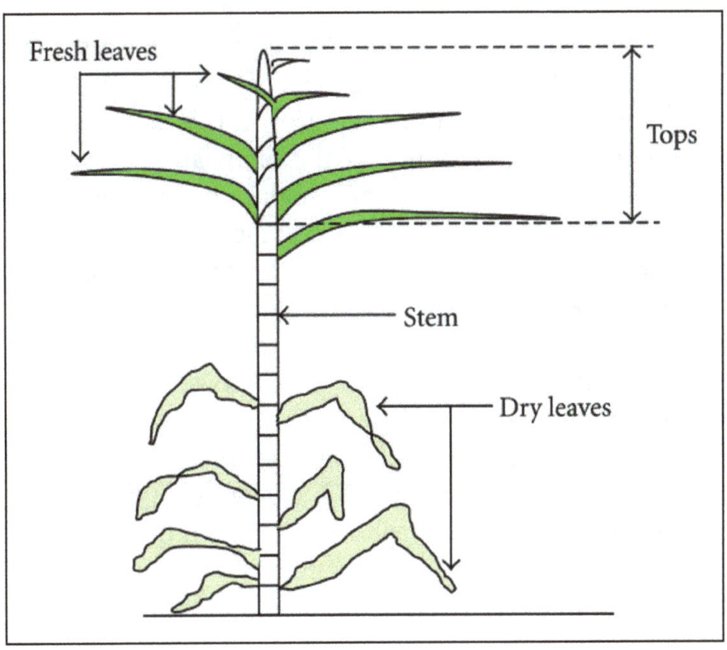

Fig: The sugarcane plant morphology.

Sugarcane is composed by stem and straw (or trash). Sugarcane stem is the material removed before the milling of cane to obtain a juice which is subsequently used for sugar (sucrose) or alcohol (ethanol) production. SB is the residue from stems after extraction of juice. SS (or trash) is composed by fresh leaves, dry leaves, and tops available before harvesting. Fresh leaves are green and yellow in color, tops are the part of cane plant between the top end and the last stalk node, and dry leaves are normally in brownish color.

Potential uses of the leaves include:

- As a fuel for direct combustion.
- As a raw material for conversion by pyrolysis to char, oil, and/or gas.
- As a raw material for conversion by gasification and synthesis to methanol.

Potential uses for the tops include:

- As a ruminant feed, either fresh or dried.
- As a substrate for anaerobic fermentation to methane.
- After reduction in water content, for the energy uses listed for cane trash.

SB and SS are normally burnt in the open agricultural field after the harvesting of the crop, or in some cases, used as an untapped source of simple sugars that can be utilized for the alcohol production.

Characterizing the sugarcane stalk, SS and SB, following observations can be interpreted: the sucrose accumulation is greater at the base of the stem, and the amount of reducing sugars and cellulose contents is superior in the tops; the length of the stem depends on some factors like the variety of the plant and the cultural management given, so that an adult stem may have from less than two meters to over four in size, affecting the length and the number of internodes; the diameter of the stem also varies, oscillating in its middle part from 250 to 350 m; the color depends on the chlorophyll content and that of anthocyanins as well as on aspects of agronomy; there is a large variation in the moisture content of the sugarcane material, varying from 13.5% (in dry leaves) up to 82.3% (in the tops); the values of ash, fixed carbon, and volatile matter have little variation among the three components of the straw, with a lower amount of ash for the bagasse; all material present practically the same composition in carbon (~45%), hydrogen (~6%), nitrogen (0.5–1%), oxygen (~43%), and sulfur (~0.1%); mineral composition for alkalis and phosphorus shows some variation among the three components of the SS, indicating that its content grows from the dry leaves to the tops, and is quite higher than SB.

Physical and Chemical Compositions of Sugarcane

Physically, sugarcane is constituted by four major fractions, whose relative magnitude depends on the sugar agroindustrial process: fiber, nonsoluble solids, soluble solids, and water. The fiber is composed of the whole organic solid fraction, originally found in the cane's stem, and characterized by its marked heterogeneity. The nonsoluble solids, or the fraction that cannot be dissolved in water, are constituted mainly by inorganic substances (rocks, soil, and extraneous materials), and it is greatly influenced by the conditions of the agricultural cane processing, types of cutting, and harvesting. Soluble solids fractions that can be dissolved in water are composed primarily of sucrose as well as other chemical components such as waxes, in a smaller proportion. SB or SS which are the focus of 2G ethanol production are lignocellulosic materials chemically composed by cellulose, hemicelluloses, and lignin.

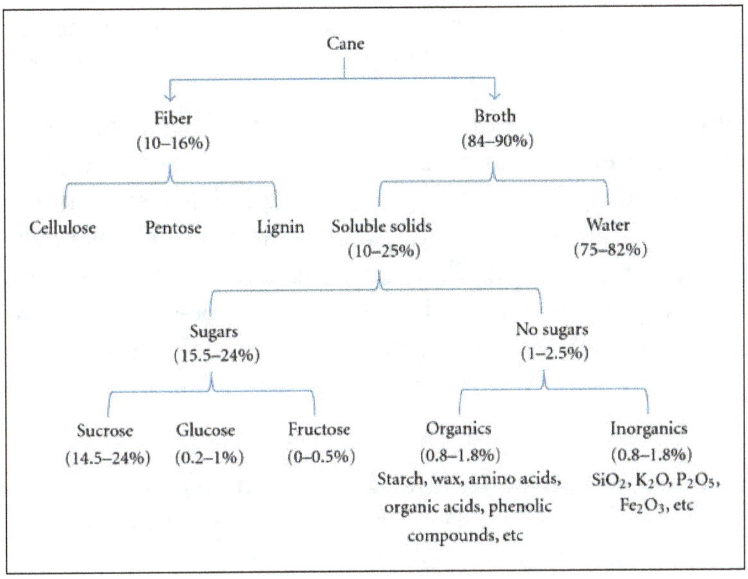

Fig: General composition of Sugarcane.

Cellulose is a linear polymer of glucose units linked by β (1→4)-glycosidic bonds, forming cellobiose that is repeated several times in its chain. This cellulosic fraction can be converted into glucose by enzymatic hydrolysis, using cellulases, or by chemical way, using acids like sulfuric acid, that subsequently can be fermented to ethanol. Hemicellulose is a heteropolysaccharide composed by hexoses (D-glucose, D-galactose, and D-mannose), pentoses (D-xylose, L-arabinose), acetic acid, D-glucuronic acid and 4-O-methyl-D-glucuronic acid units. The hemicelluloses are classified basically according to the sugars that are present in the main chain of polymer: xylan, glucomannan, and galactan. The hemicellulose differs substantially of cellulose to be amorphous, which make it easier to be hydrolyzed than cellulose. The hemicellulosic fraction can be removed from lignocellulosic materials by some type of pretreatment, like acid or hydrothermal hydrolysis, and liberating sugars, mainly xylose, that subsequently can also be fermented to ethanol.

Lignin is a complex aromatic macromolecule formed by radical polymerization of three phenyl-propane alcohols, namely *p*-coumarilic, coniferilic, and synapilic. In the plant cell wall, lignin and hemicelluloses involve the cellulose elementary fibrils, providing protection against chemical and/or biological degradation. The content of lignin and its distribution are the responsible factors for the recalcitrance of lignocellulosic materials to enzymatic hydrolysis, limiting the accessibility of enzyme, and therefore, the process of delignification can improve the conversion rates of enzymatic hydrolysis. The lignin is primarily used as a fuel, but it can be chemically modified to be used as chelating agent, for removal of heavy metals from wastewater, or as precursor material for production of add-value products as activated carbon, surfactants, and adhesives.

SB of the Brazilian territory is quantitatively composed by 38.4–45.5% cellulose, 22.7–27.0% hemicellulose, and 19.1–32.4% lignin. Nonstructural components of biomass, namely, ashes (1.0–2.8%) and extractives (4.6–9.1%) are the other substances that compose the chemical compositional of bagasse.

The ash content of SB is lower than the other crop residues like rice straw and wheat straw (with approximately 17.5 and 11.0% of this compound, resp.). SB is also considered a rich solar energy reservoir due to its high yields and annual regeneration capacity (about 80 t/ha) in comparison with agricultural residues like wheat, grasses, and trees (1, 2, and 20 t/ha, resp.). The bagasse can also be used as a raw material for cultivation of microorganisms and for bioconversion process for the production of industrial enzymes, xylitol, and ethanol production. Due to these advantages the bagasse is considered not only a sub-product of sugar industry, but also a co-product of high added value.

The fact that chemical composition varies for the same type of material did not cause surprise because the major fractions of lignocellulosic materials depend on many factors including plant genetics, growth environment, and processing conditions as well as methods employed for the compositional analysis. It is impossible to compare the composition of samples of different origins, performed by different laboratories that do not use the same methods.

The large variation in the values of chemical components also is observed for the SS. Generally it is composed of 33.3–36.1% cellulose, 18.4–28.9% hemicellulose, and 25.8–40.7% lignin. Ashes (2.1–11.7%) and extractives (5.3–11.5%) are also present in SS. The amount of straw from sugarcane harvesting depends on several factors such as: harvesting system, topping, height, cane variety, age of crop (stage of cut), climate, soil, and others. For example, when mechanically harvested, and depending on the harvesting technology applied, the range of straw that is collected and transported to the mill together with the stems is from 24% to 95% of the total trash available.

The growing concern over the shortage of oil reserves, together with the need to preserve the environment, is the main drivers of the search for viable alternative renewable sources for the production of sustainable fuels. Furthermore, since the transportation sector accounts for about 16% of global greenhouse gas (GHG) emissions and about 27% of total US GHG emission, the replacement of oil-derived fuels by biofuels could contribute to the reduction of environmental impacts as well as provide socio-economic benefits. There is already a consensus that a shift of the global energy scenario towards increasing the share of renewable sources is necessary and that ethanol will undoubtedly be an important biofuel. The need to produce inexpensive renewable fuels to substitute fossil fuels is revealed in the political agendas of several countries. Therefore, the future energy economy is likely to be based on a wide range of alternative energy platforms. Among the potential bioenergy sources, the lignocellulosic biomass has been identified as an important feedstock for the production of biofuels and other value-added products, thus contributing to the global energy supply. In this context, Brazil is notable in that the country has made considerable progress towards replacing fossil fuels by renewable ethanol from sugarcane. Moreover, Brazil is the world's largest producer of sugarcane and together with US leads the global production of bioethanol. In the 2014/15 harvest, it is estimated that more than 640 million tons of sugarcane will be processed by the Brazilian sugar-alcohol mills, which will result in an ethanol production of approximately 28 billion liters and a sugar production of about 36 million tons.

In the sugarcane industry, large amounts of lignocellulosic residues (bagasse and trash) are generated during the production of ethanol and sugar. After the sugarcane harvest, the trash remaining consists of three main components, namely, dry leaves, green leaves, and tops. The tops are the segments of the sugarcane plant between the upper end and the last stalk node with attached green leaves, while the dry leaves are known as straw. The milling of the sugarcane to extract the juice generates the bagasse. In this process, every ton of sugarcane processed generates 140 kg of bagasse and 140 kg of trash, on a dry basis (db). In contemporary facilities, the bagasse is currently used as fuel, that is, it is burned in boilers to meet the energy demands of the industrial plant, with the surplus electric energy being exported to the grid. The sugarcane trash was previously burned prior to the harvest in order to facilitate the harvest procedure, but is now mostly left in the field for agricultural purposes as fertilizer and for pest control. There is currently an ongoing progressive shift in the sugarcane harvesting method, from manual harvesting of burned sugarcane to mechanical harvesting of unburned sugarcane, with the trash remaining on the ground. This increasing amount of lignocellulosic material being left on the ground could be partially recovered and used for energy generation in the mills, thus improving the overall energy balance.

The second-generation (2G) ethanol produced from lignocellulosic biomass has been considered to be the biofuel with the greatest potential to replace oil-based fuels. However, 2G technology is not as mature as that of conventional first-generation (1G) ethanol, and the process is still less economically feasible. In energy terms, it is known that the energy content (MJ/kg) of sugarcane trash is similar to that of the bagasse, with the two components accounting for approximately two thirds of the total energy content of sugarcane. Despite the fact that the trash is currently considered the main unexploited energy reserve in the sugarcane agro-industry, its most likely use in the short to medium term will be as boiler fuel. Nonetheless, considering the amounts of residues generated and that all the cellulose present in these materials could be converted into ethanol, the use of the whole sugarcane lignocellulosic biomass (including trash and bagasse) could greatly increase ethanol production without the need for expansion of cultivated areas. It is worth mentioning that there is still no consensus about the optimal amount of trash that should be left on the ground. It has been observed that 65% of the residue from sugarcane harvesting can be removed from the fields as dry leaves (straw), without causing negative agricultural effects.

There have been several studies of 2G ethanol production processes using sugarcane bagasse. In one study, evaluation was made of the bagasse from 115 varieties of sugarcane developed by classical and precision breeding technologies, in terms of fiber composition and fermentable sugar yields after dilute acid pretreatment and enzymatic hydrolysis. It was suggested that variety selection could contribute to the development of a cost-effective pretreatment and saccharification process.

Soybean Biofuel

Soybean (Glycine max) is a major crop throughout much of North America, South America and Asia. The United States is the world's greatest producer, producing approximately 32% of the world's soybeans followed by Brazil with 28%. Origins of soybean are in Southeast Asia with first domestication reported in the 11th century BC in China. First planted in the U.S. in 1765, soybeans spread to the Corn Belt by the mid-1800s with major acreage not seen until the 1920s when it was used mainly as a forage crop. Major U.S. expansion as an oilseed crop began in the 1940s (Gibson and Benson 2005). Soybeans contain approximately 18 to 20% oil compared to other oilseed crops such as canola (40%) and sunflower (43%). At 48 lbs per bushel soybean meal remains a major product from soybeans used for animal feed and human food. Soybean acreage is much greater than other oilseed crops leading to substantial soybean oil production and its availability as a biofuel feedstock.

Current Potential for Use as a Biofuel

Soybean oil is currently a major feedstock for production of biodiesel. The most common method of biodiesel production is a reaction of vegetable oils or animal fats with methanol or ethanol in the presence of sodium hydroxide (which acts as a catalyst). The transesterification reaction yields methyl or ethyl esters (biodiesel) and a byproduct of glycerin. Numerous studies between 1980 and 2000 have shown the use of straight vegetable oil including soybean oil to cause carbon deposits and shorten engine life. Biodiesel use in diesel engines does not have similar negative effects. Use of soybean oil for biodiesel was greatly influenced by promotion from U.S. soybean farmers through the United Soybean Board (USB) and subsequent creation of the National Biodiesel Board.

Biology and Adaptation

Soybean (Glycine max) is a cool season legume which can be grown from south to north throughout much of the eastern half of the U.S. Soybeans and other legumes have a unique relationship with a bacteria bradyrhizobium species, will colonize on soybean roots forming a nodule. The two species form a symbiotic relationship where the soybean plant provides nutrition and the bacteria fixes nitrogen from the air. This relationship reduces the need for supplemental nitrogen fertilizer in soybean production.

Soybeans flower in response to day length and temperature. Varieties grown in the United States are divided into 13 maturity groups from maturity group 000 being the earliest and adapted to northern regions of Minnesota and southern Canada, to maturity group X adapted to southern regions such as south Texas. The earlier varieties bloom when days are long and nights are short, while the later-maturing varieties bloom under relatively shorter days and longer nights. Summer days are longer at northern latitudes, where early maturing varieties will initiate flowering when days are longer. Maturity groups develop differently and knowing the growth habit of different maturity groups can help with the crop management.

Production and Agronomic Information

Through much of the upper Midwest soybeans are planted in April to June and harvested in September to November. Soybeans are well adapted to grow in soils similar to corn production. In many cases soybeans are grown in rotation with corn or wheat to break insect, weed, and disease cycles. Nutrient requirements are generally less for soybeans than other crops with major nutrient requirements nitrogen, phosphorous, and potassium and where much of the nitrogen is gained through a relationship with bacteria. A soil pH in the range of 5.5 to 7.0 will enhance nutrient availability and soybean growth. Weed control is necessary to achieve optimal yields and use of biotech seeds has eased the ability to control weeds during the growing season. Currently in the U.S. over 90% of soybeans planted are herbicide resistant. Many insects and diseases are common in soybeans grown in the upper Midwest. The most damaging pest to soybeans is soybean cyst nematode, a soil born parasitic roundworm feeding on soybean roots. Insect pests include: bean leaf beetle, soybean aphid, green clover worm, and spider mites. Soybean harvest begins after 95% leaf senescence when beans are 12 to 18% moisture.

Potential Biofuel Yields

Current U.S. production of soybeans in 2009 was 3.4 billion bushels from 77.4 million acres. Average yield per acre for the U.S. was 44 bushels per acre (National Ag Statistics Service). One bushel of soybeans can yield 1.5 gallons of biodiesel. Using all U.S. soybeans for biodiesel could produce 5.1 billion gallons biofuel. Using all soybean production for biodiesel has not been proposed and is not realistic. In 2009 biodiesel production was 700 million gallons with a production capacity of 1.83 billion gallons. Based on a yield of 44 bushels per acre, an acre of soybeans could yield 66 gallons of biodiesel compared to 69 gallons for a 1300 lb per acre canola yield, 84 gallons for sunflower and over 600 gallons for palm oil.

Production Challenges

Soybean production generally compliments corn production in the upper Midwest. Both corn and soybeans enjoy a long history of production on millions of acres in the upper Midwest. This history has led to a large infrastructure of equipment, storage, rail, barge and truck transportation. Soybeans like many crops face insect and disease pests along with weather related challenges. An emerging disease has gained much attention in recent years. Soybean rust, a fungal disease native to Asia has spread to the soybean fields of South America and finally to U.S. soybeans. Rust control is expensive requiring fungicide applications and yield damage can be extreme.

Estimated Production Costs

Production costs will vary depending on location, cropping systems, and fluctuation in price of energy. Major expenditures in soybean production include; Planting, harvesting, seed and pesticides. An example of a Nebraska rainfed budget for no-till soybeans for 2010 lists $115 per acre for field operations, materials and services. When including overhead costs for land, insurance, etc. the total is approximately $200 per acre. Total costs for irrigated soybeans are around $400 per acre. Biodiesel profitability is extremely variable and based on the continuously changing prices biodiesel, soybean oil, co-product glycerin, methanol and natural gas. Price of soybean oil feedstock is one of the driving factors in profitability.

Environmental and Sustainability Issues

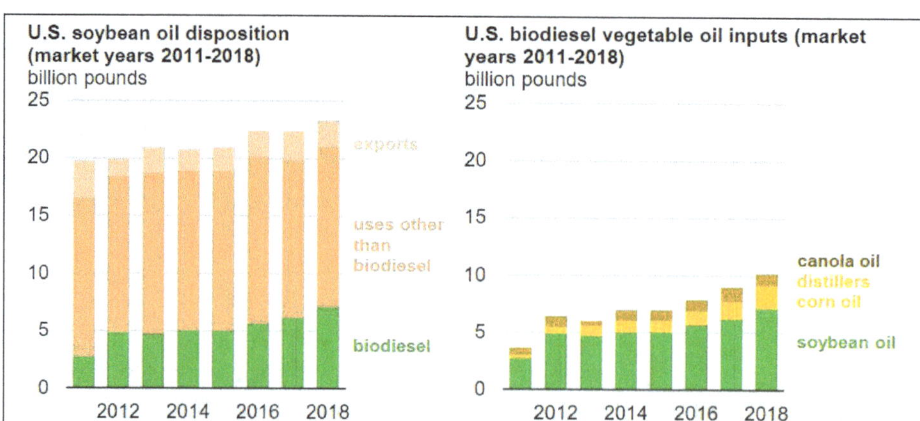

Soybeans use for biodiesel production capacity grew from zero to over a billion gallons per year in the past two decades 90s and 00s. Through that time biodiesel production rose and fell depending on the price of feedstock, price of petroleum oil, and federal and state subsidies provided to the industry. One major challenge for soybeans are competing uses for soybean oil. Soybean oil is used for many human food products, cooking oil and numerous industrial applications. Soybeans account for 80 percent or more of the edible fats and oils consumed in the U.S. Competition with other uses has caused price spikes in the soybean oil market challenging the profitability of soybean biodiesel. A life cycle analysis of biodiesel done by the USDA reports the fossil energy ratio of biodiesel to be 3.2 units of energy out for each unit of fossil energy used. In other words, biodiesel yields 3.2 units of energy for every unit of fossil energy consumed compared to petroleum diesel which has a fossil energy ratio of ~ 0.84.

Camelina Biofuel

Camelina [Camelina sativa (L.) Crantz], also known as false flax or gold-of-pleasure, is a broadleaf oilseed flowering plant of the Brassicaceae (mustard) family that grows optimally in temperate climates. Other more common members of this family include important food crops such as broccoli, Brussels sprouts, cabbage, cauliflower, collards, kales, kohlrabi, radish, rapeseed/canola, rutabaga, turnip and various mustards. Camelina can be grown in a variety of climatic and soil conditions as a spring or summer annual or as a biannual winter crop. Cultivated in Europe sporadically since the Bronze Age, camelina has several beneficial agronomic attributes: a short growing season (85–100 days), compatibility with existing farm practices and tolerance of cold weather, drought, semiarid conditions, and low-fertility or saline soils. Camelina also has lower water, pesticide, and fertilizer requirements than other traditional commodity oilseed crops (rapeseed/canola, soybean, and sunflower, for example). Moreover, camelina is well adapted to the more northerly regions of North America, Europe, and Asia. As such, it may serve in a rotational cycle in the Northern Hemisphere where winter wheat (Triticum spp.) is typically planted, thus facilitating disruption of undesirable weed and pest cycles. Some estimates indicate that the northern U.S. state of Montana alone could support between 0.8 and 1.2 million hectares of camelina per year.

Camelina Oil and Meal

Camelina yields anywhere from 336 to 2240 kg of seeds per hectare at maturity with the lipid content of individual seeds ranging between 35 and 45 weight percent (wt%). The resulting yield of camelina oil is thus calculated to be between 106 and 907 liters per hectare, which as seen in Table below at the higher end of the range. It is higher than soybean and sunflower oils but lower than rapeseed oil. The wide variability in yield is attributable to such factors as the climate and conditions at the growing location as well as by the amount of agricultural inputs applied to the crop and the presence of pests, weeds and disease. With further genetic modification, the yield of camelina may approach that of rapeseed.

Table: Comparison of yields from several oilseeds.

	Camelina	Rapeseed	Soybean	Sunflower
Seed yield (T/ha)	0.90–2.24	2.68–3.39	2.14–2.84	1.44–1.70
Oil content (wt%)	35–45	40–44	18–22	39–49
Oil yield (L/ha)	106–907	965–1342	347–562	505–750

α-Linolenic acid generally comprises between 32 and 40 wt% of the fatty acid composition of camelina oil. Other fatty acids in quantities above 10 wt% include linoleic, oleic, and 11-eicosenoic acids. The acid value of crude camelina oil extracted either by solvent (hexane) or cold press is typically between 1 and 5 mg KOH/g of oil, which is considerably lower than many other crude plant oils. Camelina oil has been used for numerous applications including directly as an experimental fuel for diesel transport engines, as culinary oil, and as a biological feedstock for the experimental production of long chain wax esters for potential cosmetic and lubricant applications. Camelina oil may also be used as an industrial source of α-linolenic acid, as it has a comparatively high content of this constituent versus the commodity oils.

Fatty acid	Carbon number	Leonard	Moser & Vaughn	Zuhr & Matthaus	Frohlich & Rice
Palmitic	16.0	5.3	6.8	5.4	5.4
Stearic	18.0	2.5	2.7	2.5	2.6
Oleic	18.1	12.6	18.6	14.9	14.3
Linoleic	18.2	15.6	19.6	15.2	14.3
α-Linolenic	18.3	37.5	32.6	36.8	38.4
Arachidic	20.0	1.2	1.5	1.3	1.4
11-Eicosenoic	20.1	15.5	12.4	15.5	16.8
Eicosadienoic	20.2	2.0	1.3	1.9	0.2
Eicosatrienoic	20.3	1.7	0.8	1.6	2.9
Behenic	22.0	0.3	0.2	0.3	3.7
Erucic	22.1	2.9	2.3	2.8	
Others	-	2.9	1.2	1.8	

Extraction of oil from camelina seeds by mechanical expeller yields a meal that consists of approximately 10% residual oil, 45% crude protein, 13% fibers, 5% minerals, and other minor constituents such as glucosinolates and vitamins. The concentration of glucosinolates in dry camelina seeds ranges from 13 to 36 lmol/g. When ingested in sufficient quantity, glucosinolates cause deleterious effects in animals such as reduced palatability as well as decreased growth and production. Consequently, camelina meal cannot exceed 10 wt% of the total food ration given to feedlot beef cattle and broiler chickens in the United States. However, even a 10% ceiling represents significant market potential where large numbers of animals are raised for human consumption, thus representing an important revenue stream for camelina meal.

Camelina Oil as Feedstock for Biodiesel

Biodiesel can be prepared from a wide variety of lipids. Advantages of biodiesel include positive energy balance, domestic and renewable origin, biodegradability, high flash point, excellent lubricity, low or no sulfur content, lower overall exhaust emissions versus petrodiesel and miscibility with petrodiesel at all blend ratios. Poor economics and insufficient supply of feedstock represent disadvantages of biodiesel along with relatively poor cold flow properties and oxidative stability as compared with petrodiesel, as well as dilution of engine oil and elevated nitrogen oxides in exhaust emissions. Camelina is particularly attractive as an alternative feedstock for biodiesel production as a result of its low cost versus commodity oils coupled with its potential to significantly enhance domestic feedstock availability.

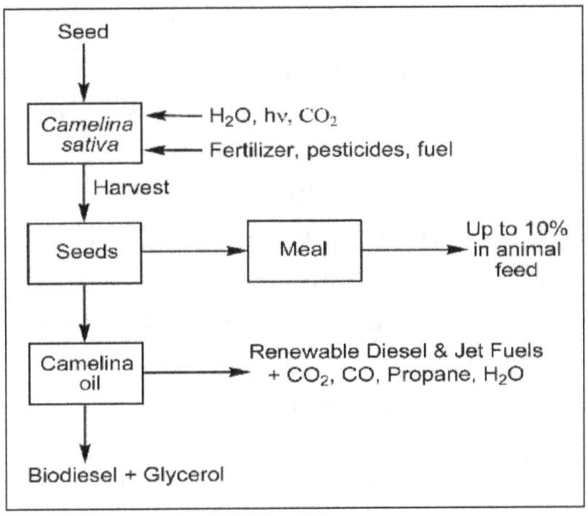

Biodiesel is classically prepared by transesterification of lipids in the presence of an homogenous alkali catalyst and excess methanol at elevated (60°C) temperature. Figure above illustrates a typical schematic for industrial production of biodiesel. Crude feedstocks that have low free fatty acid contents (less than 3.0 wt%) can be directly transesterified without pretreatment, thereby eliminating a costly pretreatment step. Camelina is one such crude oil and has been successfully converted to biodiesel (fatty acid methyl esters) by the classic method as well as with heterogeneous metal oxide catalysts both with and without microwave irradiation and at non-catalytic sub- and supercritical conditions employing co-solvents with methanol. The fuel properties (cold flow properties, oxidative stability, kinematic viscosity, cetane number, etc.) of camelina-based biodiesel are similar to those of biodiesel prepared from soybean oil, thus indicating its acceptability for use as biodiesel. Additionally, fatty acid ethyl esters have been prepared from camelina oil and along with methyl esters were evaluated as blend components in petrodiesel (a15 ppm S). As was the case with the neat esters, camelina-based biodiesel blends in petrodiesel exhibited fuel properties comparable to the corresponding soybean-based blends.

Renewable Diesel Fuels from Lipids

Renewable diesel is defined by the U.S. Internal Revenue Code (IRC) as a fuel produced from biological materials by thermal depolymerization that meets the fuel specification requirements of ASTM D975 (petroleum diesel fuel) or ASTM D396 (home heating oil). The industrial reaction by which lipids are converted to renewable diesel is referred to as hydrodeoxygenation (hydrotreatment) and normally involves elevated pressure (40–150 atm) and temperature (350–450°C) in the presence of a heterogeneous catalyst and hydrogen to achieve sufficient conversion. All or a fraction of the linear alkanes (paraffins) resulting from hydrotreatment can be isomerized in a second stage to improve cold flow properties, although isomerized hydrocarbons have significantly lower cetane numbers than their linear counterparts. Renewable diesel is composed of C15–C18 paraffins with a boiling range (260–320°C) within that of traditional petrodiesel (200–350°C).

The chemical composition and fuel properties of renewable diesel are different from biodiesel and in many cases are closer to that of petrodiesel. While the renewability aspect of renewable diesel is retained versus biodiesel, advantages such as biodegradability, positive energy balance, excellent lubricity, and high flash point are sacrificed. However, disadvantages such as poor oxidative stability, cold flow properties, and energy density as well as elevated NOx exhaust emissions versus petrodiesel are eliminated if renewable diesel is prepared instead of biodiesel. Furthermore, renewable diesel can be produced using existing petroleum oil refinery capacity and is compatible with the current diesel fuel distribution infrastructure.

Table: Comparison of biodiesel to renewable diesel with petrodiesel as a reference.

	Renewable Diesel	Biodiesel
Chemical Structure	Hydrocarbons	FAME
Catalyst	Heterogeneous	Homogenous
Requirements	H_2, Pressure, Heat	Methanol, Heat
Feedstock FFA problematic?	No	Yes
Byproducts?	Propane	Glycerol
Positive energy balance?	No	Yes
Renewable feedstock?	Yes	Yes
Biodegradable fuel?	No	Yes
Relative to petrodiesel:		
Density	Similar	Higher
Viscosity	Similar	Higher
Sulfur content	Lower	Lower
Flash point	Similar	Higher
Lubricity	Similar	Better
Cetane number	Higher[a]	Higher
NOx exhaust emissions	Similar	Higher
Cold flow properties	Similar[b]	Lower
Oxidative stability	Similar	Lower
Energy density	Similar	Lower

[a] Linear paraffins.

[b] Isomerized (branched) paraffins.

Camelina Oil as Feedstock for Renewable Jet Fuel

Production of renewable jet fuels from vegetable oils initially requires hydrodeoxygenation similar to renewable diesel fuel. However, isomerization to branched paraffins to improve cold flow properties is critically important in the case of jet fuel, as linear paraffins freeze at the temperatures encountered at the operational altitudes of most commercial aircraft. Consequently, ASTM D1655 requires that aviation turbine (jet) fuels have freezing points no higher than −40°C (−47°C for JP8). Selective catalytic cracking of the paraffins to shorter-chain hydrocarbons (C7–C18) is also performed, as jet fuels have a somewhat lower boiling range than diesel fuel. With existing catalyst and reactor technologies, selective catalytic cracking and isomerization can be accomplished in a single stage. Lastly, aromatics can be introduced either by blending renewable hydrocarbons with petroleum jet fuel or by conversion of a portion of the bio based paraffins into branched aromatics by ring-closing aromatization. Because aromatization would add a third stage to the production of renewable jet fuel, blending with petroleum jet fuel obviously represents a more viable option from economic and process standpoints.

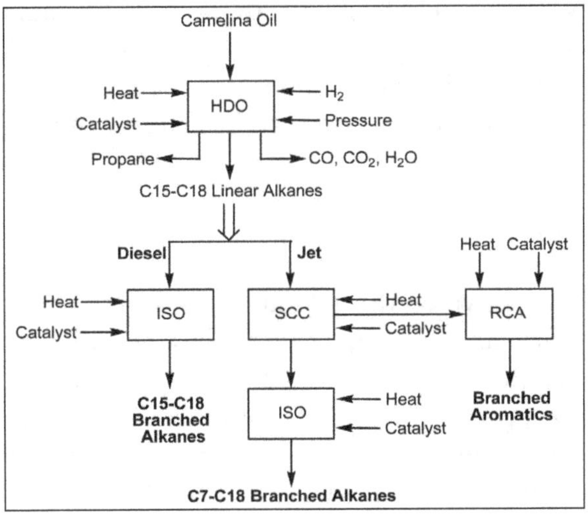

Simplified scheme for industrial production of renewable diesel and jet fuels from camelina oil; HDO = hydrodeoxygenation; ISO = isomerization; SCC = selective catalytic cracking; RCA = ring-closing aromatization. Renewable diesel = mixture of C15–C18 branched and linear alkanes; Renewable jet fuel = mixture of branched aromatics and C7–C9 branched alkanes.

Rapeseed

Rapeseed is one of the most important oils used for the production of biodiesel. In Europe rapeseed is one of the most common oil used for production of Biodiesel. Rapeseed is also known as Brassica Napus.

Scientific Classification:

- Kingdom: Plantae.
- Division: Magnoliophyta.
- Class: Magnoliopsida.
- Order: Capparales.
- Family: Brassicaceae.
- Genus: Brassica.
- Species: B. napus.
- Binomial name: Brassica napus L.

Rapeseed (Brassica napus), also known as Rape, Oilseed Rape, Rapa, Rapaseed and (one particular variety) Canola, is a bright yellow flowering member of the family Brassicaceae (mustard or cabbage family). The name is derived through Old English from a term for turnip, rapum (see Brassica napobrassica, which may be considered a variety of Brassica napus). Some botanists include the closely related Brassica campestris within B. napus.

Cultivation and Uses

Rapeseed is very widely cultivated throughout the world for the production of animal feed, vegetable oil for human consumption, and biodiesel; leading producers include the European Union, Canada, the United States, Australia, China and India. In India, it is grown on 13% of cropped land. According to the United States Department of Agriculture, rapeseed was the third leading source of vegetable oil in the world in 2000, after soybean and oil palm, and also the world's second leading source of protein meal, although only one-fifth of the production of the leading soybean meal. World production is growing rapidly, with FAO reporting that 36 million tons of rapeseed was produced in the 2003-4 season, and 46 million tons in 2004-5. In Europe, rapeseed is primarily cultivated for animal feed (due to its very high lipid and medium protein content), and is a leading option for Europeans to avoid importation of GMO products.

The rapeseed is the valuable, harvested component of the crop. The crop is also grown as a winter-cover crop. It provides good coverage of the soil in winter, and limits nitrogen run-off. The plant is ploughed back in the soil or used as bedding. On some ecological or organic operations, livestock such as sheep or cattle are allowed to graze on the plants.

Processing of rapeseed for oil production provides rapeseed animal meal as a by-product. The by-product is a high-protein animal feed, competitive with soya. The feed is mostly employed for cattle feeding, but also for pigs and chickens (though less valuable for these). The meal has a very low content of the glucosinolates responsible for metabolism disruption in cattle and pigs. Rapeseed "oil cake" is also used as a fertilizer in China, and may be used for ornamentals, such as Bonsai, as well.

Biodiesel

Rapeseed oil is used in the manufacture of biodiesel for powering motor vehicles. Biodiesel may be used in pure form in newer engines without engine damage, and is frequently combined with standard diesel in ratios varying from 2% to 20% biodiesel. Formerly, due to the costs of growing, crushing, and refining rapeseed biodiesel, rapeseed derived biodiesel cost more to produce than standard diesel fuel. Rapeseed oil is the preferred oil stock for biodiesel production in most of Europe, partly because rapeseed produces more oil per unit of land area as compared to other oil sources, such as soy beans.

Process for Obtaining Bio-Diesel

Transesterification is the best way to obtain biodiesel because it is well-known and cheap process which gives fewer problems for the engines than other methods. In a standard process of production biodiesel from rapeseed oil there are following process steps i.e. esterification of rapeseed oil, separation of esterification products, methanol distillation and purification of the ester. The main stage of the process is based on the transesterification reaction of rapeseed oil with an alcohol (methanol, ethanol) which results in formation of esters of alcohols and glycerol. The reaction is reversible due to formation of water, which is responsible for shifting the equilibrium towards the reagents. In order to move the chemical equilibrium towards the ester, an excess of alcohol is used. The transesterification involves three consecutive and reversible reactions. Each molecule of alcohol makes with the residue of fatty acid a distinct monoester molecule. If the triglyceride contains three different fatty acids in its molecule, a mixture of rapeseed oil esters is obtained.

There are used various chemical catalysts in the transesterification reaction i.e. acids, alkalis and enzymes. Some of them are the most effective i.e. alkaline catalysts and their methoxides. Only anhydrous reactants should be used, because the water decomposes catalyst and leads the reaction to the side of the reactants. The amount of the catalyst depends on its type, quality of substrates, reaction time and temperature. Its value varies from 0.2% to 2% by mass relative to the weight of oil. Unrefined oil requires more catalyst. Free fatty acids content in oil shouldn't exceed 0.5-1%. The higher the water content in methanol, the greater the consumption of catalyst and higher contents of free fatty acids as well as soaps. Forming of soaps reduces the activity of the catalyst and increases the viscosity, so the separation of the glycerol is more difficult.

In the majority of the biodiesel production processes, methanol is used, mainly due to the cost and its physical and chemical advantages (polar, shortest chain alcohol and the reactivity is easier). For the basic-catalysed process, the optimal ratio is 6:1 (alcohol: oil). Applying alkaline catalysts in the reaction allow to carry out the transesterification process at room temperature. The reaction time is very diverse and can vary from several minutes to several hours (typically about 30-60 minutes). Mixing is important because of the lack of mutual solubility of the substrates. Appropriately intensive mixing of reactants increases the contact area. Using the refined oils with a higher purity as well as pure and anhydrous methanol, allows to obtain biodiesel with a high content of methyl ester (96.5- 99%).

In the transesterification reaction, the oil is converted to FAME (Fatty Acid Methyl Ester). The flow, which goes out from the reactor, should be purified. First, the methanol is recovered (94%) and then residue is carried to separation of glycerin. Next, the obtained biodiesel is purified to commit the standards of quality.

The primary by-products generated during the production of biodiesel include: oilseed meal, straw and glycerol. Oilseed meal is produced in the process of pressing oil from rapeseeds. It accounts for 70% by weight, and oil represents the remaining 30%. Oilseed meal can be used as a component of feed or as a component of the binder for the production of smokeless fuel, additive to fuel briquettes and as a basic component of mat to cover mine dumps. Straw is formed at the beginning of the production cycle. Part of straw in the form of chaff can be plowed already on the field and is a valuable organic fertilizer. Properly prepared straw can also be used as a component of the roughage.

Straw is also a valuable energy source, because its calorific value is about 30-40% of the calorific value of coal and the ash formed of its combustion can be used as a mineral fertilizer. Glycerol is a byproduct formed in the esterification of rapeseed oil. In this process also are created soaps and free fatty acids. Glycerol phase constituting concentrated glycerol (about 80% solution) may be transferred to facilities specializing in treatment of glycerol (e.g. cosmetic, pharmaceutical). The cost of obtaining the crude glycerol during esterification of vegetable oils is lower than the cost of traditional methods that had been previously used in splitting of fats.

Glycerol from vegetable oils is used both for cosmetic purposes and pharmacology. More often it is an ecological material for the manufacture of certain foods and it is used for making explosives. Using own purification of glycerol, receives an additional by-product in the form of valuable mineral fertilizer. In the case of small factories, where it is uneconomic to build installation of treatment and it is not organized buying glycerol, it can be disposed of by diluting with water in a ratio of 1:100 and added to the slurry.

Cellulosic Biofuel

Ethanol is an alcohol that can be created from a wide variety of plant materials and feedstocks and is used in liquid from to fuel in motor vehicles. Today, corn starch and sugarcane are the two main feedstocks used, respectively producing starch- and sugar-based ethanol. Cellulose can also be used to produce ethanol; however, doing so requires additional processes using enzymes to break down the cellulosic materials into sugars. Cellulosic materials, which provide structure to plants, comprise the stems, stalks, and leaves of plants as well as trunks of trees. Cellulose and hemicellulose, which collectively are referred to as cellulosic materials, can be broken down into sugars, which can then be fermented into ethanol. Cellulosic materials being studied for the production of biofuels includes those found in switchgrass, prairie grasses, cornhusks, wood chips, forestry materials and residues as well as other inedible agricultural plant waste. The ethanol produced from these cellulosic materials is referred to as cellulosic ethanol.

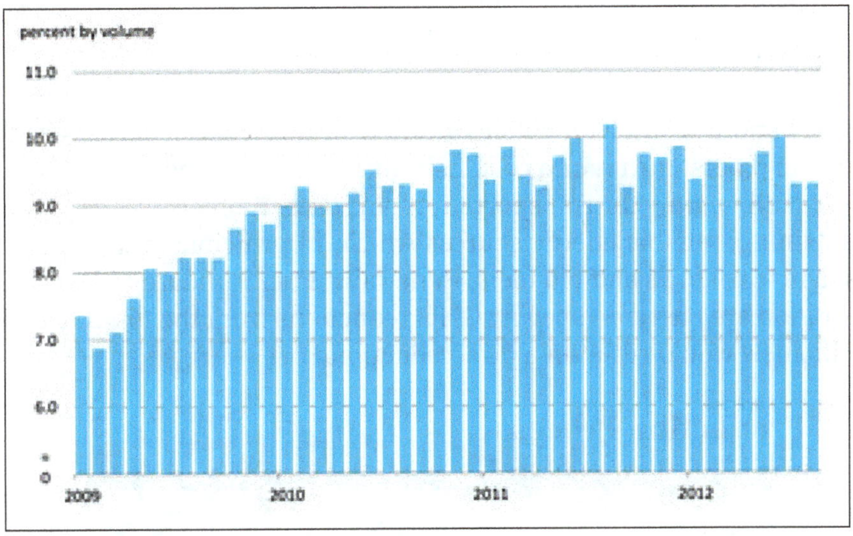

Ethanol Share of U.S. Gasoline Consumption, 2009-12.

Why Cellulosic Ethanol?

The abundance of cellulosic materials, which boils down to about 60 to 90 percent of earths biomass measured by weight, along with the fact that they are not used for food and feed, are significant reasons why cellulosic ethanol and other cellulose-based biofuels are so appealing to scientific and political scholars.

Today, corn-based ethanol comprises nearly 10 percent of U.S. motor fuel. This number has steadily increased over the past several years, jumping about 3% since 2009. However, Congress is worried about driving up the price of corn because it is used as feed for livestock and poultry, rising corn costs would subsequently raise the prices in supermarkets. Lawmakers have resorted to capping the total production of corn-based ethanol and have called for a steady increase in the use of advanced biofuels. Congress concerns are grounded in the world energy consumption which has doubled in the past 30 years, and it will almost double again in the next 30 years. By the year 2030, we will need 30 TW of average power, from which 15% will be transportation energy, mostly oil.

Currently, transportation energy is 35 million barrels of oil per day. From all the oil consumed in the world, 50% goes directly to transport. 1/4 of all oil in the world is consumed in the US, from which 65% is imported and the demand is going to rise. Cellulosic ethanol also has the hypothetical ability to provide substantial lifecycle GHG reductions compared to petroleum-based gasoline.

Using biomass for transportation fuels raises questions concerning the logistics of feedstock production such as land use and land use change, fertilizer and pesticide use, water consumption, and energy used for production and cultivation. However, grasses and trees typically require minimal labor and generally have lower fertilizer and pesticide needs and resources as opposed to other row crops such as corn. Grasses such as switchgrass require a low level of attention and are perennial crops that do not need to be re-planted for about 20 years and provide as easily accessible feedstock to produce cellulosic ethanol.

The main idea and potential benefits associated with biofuels is to extract the energy that is stored in plants, turn it into cellulosic ethanol, and replace a significant amount of the oil demand, mainly for transportation purposes. Biofuels like cellulosic ethanol are self-sustaining, reliable energy sources which, in principle, have smaller net CO_2 emissions than fossil fuels and bio fuels. Renewable Fuel Standard (RFS) goals for biofuels penetration are based on specific GHG reductions from the fossil fuel it replaces. Cellulosic Biofuels would result in a 60% reduction.

Cellulosic Production Process

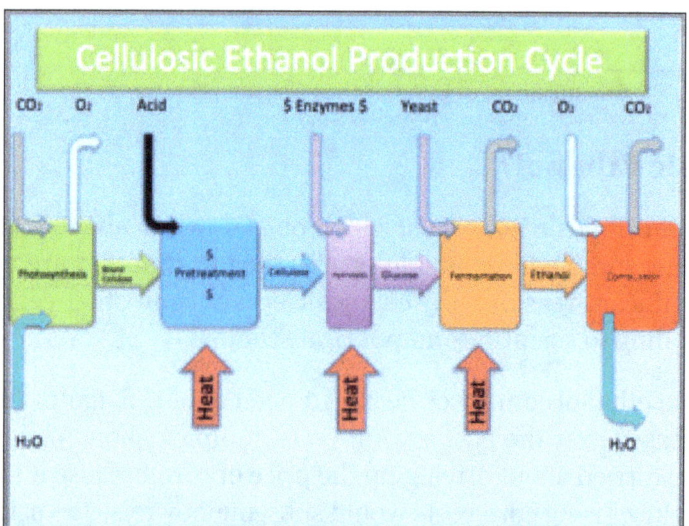

Fig: Cellulosic Ethanol Production Cycle.

The chemical make-up of ethanol is uniform across the board whether it is produced from corn, sugarcane, or cellulose; however, the differences lie in the production processes and the necessary technologies in different stages of development. While corn- and sugar-based ethanol production technologies have been produced at a commercial scale for decades, some of the technologies needed to manufacture cellulosic ethanol, an advanced biofuel, are relatively new. Today, there still no fully operational commercial-size cellulosic ethanol facilities in the United States.

The process of producing ethanol from cellulosic materials is also far more complicated than the processes employed for starch- or sugar-based ethanol. Enzymes must break up the complex cellulose-hemicellulose-lignin structure in which cellulosic materials are found before the fermentation process can begin. The cellulosic ethanol conversion process consists of two basic steps: pretreatment and fermentation. This two-step process is what is responsible for the increase in time, expense, and complexity of converting the cellulosic biomass into ethanol, relative to the procedures used to convert corn or sugarcane into ethanol.

Pretreatment

Pretreatment is necessary to prepare cellulosic materials for hydrolysis, which converts the hemicellulose and cellulose into glucose. Standard pretreatment includes a chemical pretreatment step involving acid and a physical pretreatment step such as grinding. These steps make the cellulose more accessible to the cellulases, which are the enzymes that digest cellulose and turn it into glucose. These enzymes catalyze its conversion to sugars in the successive steps and begin the breakdown of hemicellulose into glucose. Following pretreatment, the conversion of cellulose to glucose is completed using a chemical reaction called hydrolysis, normally employing enzymes secreted by certain organisms (typically fungi or bacteria) to catalyze the reaction. The pretreatment and hydrolysis process usually results in one co-product, lignin, which can be burned to generate heat or electricity. Using lignin instead of a fossil-based energy source to power the conversion process reduces cellulosic ethanol's life-cycle greenhouse gas (GHG) emissions, compared to corn-based ethanol.

Once the sugars have been derived from the cellulosic materials, they are fermented using yeast or bacteria in processes similar to those used for the corn-based ethanol production. The liquid resulting from the fermentation process contains ethanol and water; the water is removed through distillation, again similar to the corn-based ethanol process. Finding the most effective and low-cost enzymes for the pretreatment process and organisms for the fermentation process has been one of the main areas of research in the development of cellulosic ethanol. The type of feedstock and method of pretreatment both influence the amount of ethanol produced. Currently, one dry short ton of cellulosic feedstock yields about 60 gallons of ethanol. Projected yields with anticipated technological advances are as high as 100 gallons of ethanol per dry short ton of feedstock.

Cost

There are four primary factors that determine the cost of the finished product: the feedstock, chemical processing and pretreatment, refining and finishing the crude product to a usable state, and the transportation and distribution of finished fuel. However, the most significant and alarming cost can be found in the pretreatment phase. Pre-treatment is considered one of the most expensive processing steps in the bioconversion of lignocellulosic biomass, which accounts for up to 40% of the total processing cost. In addition, this trickles down and greatly affects the cost of operations such as enzymatic hydrolysis and fermentation. The catch with cellulosic ethanol is the energy required to make the enzyme. The enzyme in question is called "cellulase," which is the top-selling industrial enzyme on the market. Cellulase is typically used in a process called biostoning, which is employed to make pre-washed jeans, and also is a component in laundry detergent which is the agent responsible for removing fuzz from the cotton fibers and ultimately enhancing the brightness of the fabric. However, cellulase is not economically ideal to be used in the pretreatment process of biomass. The fungus that must be cultured and fed which makes these enzymes is costly in terms of energy. The amount of energy it takes to make the enzyme exceeds the energy produced by hydrolyzed glucose before the enzyme breaks; thus, creating the crux of the issue. There is a negative net gain of energy. The cellulose enzymes are simply too costly to use in the cellulosic ethanol process.

Benefits of Cellulose Biofuel

Cellulosic fuel crops can grow on lands that are not necessarily suitable for food crops and thereby reduce or avoid food vs. fuel competition. If grown on land that has already been cleared, cellulosic crops do not further contribute to the release of carbon to the atmosphere.

Because many cellulosic crops are perennial and roots are always present, they guard against soil erosion and better retain nitrogen fertilizer. Additionally, carbon is sequestered belowground in roots and soil organic matter because there is no further tillage after crop establishment. Most cellulosic sources require much less intensive management than do grain crops, saving the fuel and carbon dioxide costs associated with field crop operations such as planting, tillage, and weed control.

Cellulosic Biofuels Provide Environmental Benefits and Increase Ecosystem Services

Ecosystems are communities of living things and the environment in which they interact. Ecosystems are essential to life, providing innumerable and invaluable services such as clean water, food, and fiber, nutrient cycling, crop pollination, and pest suppression. Biodiversity refers to the portfolio of organisms in a natural community. Generally, the more diverse the portfolio, the greater the degree of ecosystem services provided.

Cellulosic biofuel sources can diversify agricultural landscapes by allowing farmers to grow a greater variety of crops with more complex mixtures of plant species. This increases the diversity of plants and the birds, insects, and other organisms that live in different plant communities. A mixture of native grass and tree crops can keep wildlife habitat intact and support vital ecosystem services, including those that help other crops in the landscape.

Economic Viability of Cellulosic Biofuels Depends on Location, Technology and Policy

Whether cellulosic biofuels are economically competitive and environmentally sustainable will depend on:

- Location.
- The development of new technology.
- Policies that reward environmental performance.

The economic viability of cellulosic biofuels from crop residues and perennial grasses such as switchgrass and miscanthus will differ across geographic locations. Crop yields differ across the U.S., and therefore a mix of crop types is more economically viable than a single type of crop. Likewise, costs of production – harvesting, storage, and transportation – will vary by location.

Current estimates suggest that cellulosic biofuels are likely to be more expensive to produce than grain-based biofuels. However, the advent of new technologies for harvesting, storing, and converting cellulosic sources into biofuels could make them more competitive.

Policy Implications

The U.S. Energy Policy Act of 2007 mandates that biofuels make up 22% of transportation fuel mix by the year 2022. This translates to 36 billion gallons of ethanol – 15 billion gallons of grain-based ethanol and 21 billion gallons of cellulosic ethanol. The 2008 Farm Bill offers a $1.01 subsidy per gallon for cellulosic ethanol and $45 per ton to growers of biomass feedstock for biofuels.

Policies that monetarily reward the environmental performance of alternative biofuels are needed to improve competitiveness of cellulosic biofuels relative to grain-based fuels and gasoline. Aligning energy policy and climate policy through biofuel tax credits that are inversely related to their carbon footprint can provide incentives to use high yield, low carbon cellulosic sources. Policies and incentives should also decrease reliance on exotic and invasive species, favor increased biodiversity at farm and field levels, and maintain or enhance ecosystem services.

Animal Fat Biodiesel

About one-third of the fats and oils produced in the United States are animal fats. This includes beef tallow, pork lard, and chicken fat. Animal fats are attractive feedstocks for biodiesel because their cost is substantially lower than the cost of vegetable oil. This is partly because the market for animal fat is much more limited than the market for vegetable oil, since much of the animal fat produced in the U.S. is not considered edible by humans.

Animal fat is currently added to pet food and animal feed, and used for industrial purposes such as soap making. Much of the domestic animal fat supply is exported. Animal fat feedstocks can be made into high-quality biodiesel that meets the ASTM specifications for biodiesel. However, there are some drawbacks and challenges to using animal fat feedstocks.

How Animal Fats are Processed

Waste fat from animal carcasses are removed and then made into an oil using a rendering process. Rendering consists of grinding the animal by-products to a fine consistency and cooking them until the liquid fat separates and pathogens are destroyed. The solids are usually passed through a screw press to complete the removal of the fat from the solid residue. The cooking process also removes water, which makes the fat and solid material stable against rancidity. The end products are fat, and a high-protein feed additive known as "meat and bone meal."

Fatty Acid Content of Animal Fats

Animal fats are highly saturated, which means that the fat solidifies at a relatively high temperature. Therefore, biodiesel made from animal fat has a high cloud point. For example, biodiesel made from beef tallow and pork lard has a cloud point in the range of 55°F to 60°F. B100 (pure biodiesel) made from animal fat should only be used in a very warm climate. However, animal fat biodiesel can be blended with petro-diesel. At lower blends such as B5 (a blend of 5% biodiesel with 95% petro-diesel), the high cloud point of the animal fat biodiesel does not have much effect on the cloud point of the blend.

The composition, or fatty acid profile, of various animal fats is shown in Table below. These data show how saturated the fat will be, which in turn determines how readily it will solidify as the temperature decreases.

Table: Fatty acid percentages in animal fats.

Fatty acid	Beef tallow	Pork lard	Chicken fat
Myristic 14:0	1.4 – 6.3	0.5 – 2.5	1
Palmitic 16:0	20 – 37	20 – 32	25
Palmitoleic 16:1	0.7 – 8.8	1.7 – 5	8
Stearic 18:0	6 – 40	5 – 24	6
Oleic 18:1	26 – 50	35 – 62	41
Linoleic 18:2	0.5 – 5	3 – 16	18

Beef tallow and pork lard are typically about 40% saturated (sum of myristic, palmitic and stearic acids). Chicken fat is lower at about 30-33%. For comparison, soybean oil is about 14% saturated and canola oil is only 6%. Thus, tallow and lard are usually solid at room temperature and chicken fat, while usually still liquid, is very viscous and nearly solid.

When the animal fat is made into biodiesel, the concern about solidification at lower temperatures continues. The methyl esters from the saturated fatty acids, mainly methyl stearate and methyl palmitate have high melting points.

Oxidative Stability of Animal Fat Biodiesel

In theory, the saturated fatty acids in animal fats should contribute to better oxidative stability for biodiesel. Animal fats contain very little of the polyunsaturated fatty acids, such as linoleic acid and linolenic acid, that make vegetable oils such as soybean oil and linseed oil so prone to rancidity.

However, in practice animal fat is not always more stable than vegetable oil, because vegetable oils often contain natural anti-oxidants. For example, a test of the peroxide content of lard and vegetable oil showed that the lard experienced oxidation faster than the vegetable oil.

Animal fats contain very little of the natural anti-oxidants such as vitamin E that protect vegetable oils. In some cases, used cooking oils may also contain artificial antioxidants that are added to oil to extend their life, and these artificial anti-oxidants can in turn extend the life of biodiesel made from used cooking oil.

High Cetane Number of Animal Fat Biodiesel

Animal fat feedstocks result in biodiesel with a high cetane number, which is an important quality parameter for diesel fuels. The saturated fatty acids are the source of this high cetane number and values over 60 are common. Soybean oil based biodiesel usually has a cetane number of about 48-52 and petroleum-based diesel fuel is usually between 40 and 44. When animal fat biodiesel is blended with petro-diesel, this high cetane number can help the engine start more quickly and run more quietly.

Lower Nitrous Oxide Emissions from Animal Fat Biodiesel

One of the important attributes of biodiesel is that it lowers the levels of harmful pollutants in the exhaust of diesel engines. One exception to this is oxides of nitrogen (NO_x), which is implicated in ozone and smog formation. Biodiesel tends to emit slightly more nitrous oxides than petro-diesel. Many reasons have been cited for this increase in NO_x and there is still considerable debate.

However, it has been shown that biodiesel from animal fats tends to produce a smaller increase in NOx and in some cases no increase. The primary reason for this is probably that animal fat biodiesel has a high cetane number (>60) compared with vegetable oil biodiesel (48 to 55). Higher cetane number is known to lower NO_x by lowering temperatures during the critical early part of the combustion process.

Contaminants in Animal Fat Feedstocks

Animal fats can be challenging feedstocks for biodiesel because they frequently contain contaminants that should be removed before the fuel is used in an engine.

Phospholipids, or gums, will cause insoluble precipitates when they come into contact with water. Since these precipitates will plug fuel filters, they must be removed from the fuel. Most of this material will separate with the glycerin during processing, or be removed during the purification step (water washing or ion exchange), so it is rare for these compounds to be found in the fuel. In fact, since phospholipids can de-activate exhaust after treatment devices on diesel vehicles, they are usually removed from the feedstock before it is converted to fuel. Most commonly, the gums are removed by adding water and citric or phosphoric acid to the feedstock and then separating the precipitates with a centrifuge.

Polymers that are formed naturally at the high temperatures of the rendering process can contribute to a higher viscosity in biodiesel made from animal fat. This higher viscosity can sometimes prevent animal fat biodiesel from passing the viscosity ASTM specification. If this is the case, the animal fat biodiesel can be blended with a lower-viscosity biodiesel.

An occasional problem with rendered fats is polyethylene in the fat which originates from plastic bags, ear tags, or other plastic that is mixed with the animal byproducts. Finely divided polyethylene has been found to cause cloudiness in the fuel and could plug the fuel filter. Producers who are used to working with vegetable oil may be confused by this, since with vegetable oil biodiesel, cloudiness is caused mainly by water in the fuel. If animal fat biodiesel shows cloudiness even after water is removed, this may be caused by stray polyethylene, which can be removed by passing the biodiesel through a fine filter.

High Sulfur Content of Animal Fat Biodiesel

Sulfur can sometimes be a problem for animal fat-based biodiesel. Biodiesel sold for on-highway use is only allowed to contain up to 15 ppm of sulfur. Some samples of beef tallow have been found to contain over 100 ppm of sulfur and chicken fat frequently contains a similar amount. The sulfur apparently originates from sulfur-containing amino acids associated with proteins that carry over from the rendering process.

Measurements of the sulfur levels in biodiesel produced from animal fats have shown that the sulfur level usually decreases by about half when the conversion takes place. However, the remaining sulfur can be difficult to remove. Vacuum distillation is about the only reliable technique for removing this sulfur.

Alternatively, biodiesel high in sulfur can be sold for non-highway use, such as fuel for a boiler or heater.

References

- Soybeans, bioenergy: unl.edu, Retrieved 14, June 2020

- Camelina-Camelina-sativa-L-oil-as-a-biofuels-feedstock-Golden-opportunity-or-false-hope-48856473: researchgate.net, Retrieved 09, January 2020
- Rapeseed: biodieseltechnologiesindia.com, Retrieved 30, March 2020
- Sustainability-of-cellulosic-biofuels: kbs.msu.edu, Retrieved 03, August 2020
- Animal-fats-for-biodiesel-production: extension.org, Retrieved 22, May 2020

4

Chemistry and Production of Biofuels

The production of biofuels involves various chemical reactions such as esterification and transesterification. Chemically, petrodiesel and biodiesel are very similar; the only difference is the presence of two oxygen atoms in biofuels. The topics elaborated in this chapter will help in gaining a better perspective about the chemistry and production of biofuels as well as the methods used in the production of biofuels.

Biofuel Chemistry

Biofuel Chemistry examines the combustion chemistry of compounds that constitute typical biofuels, including alcohols, ethers and esters. Biofuel is associated with only a few select chemical compounds, especially ethanol (used exclusively as a gasoline replacement in spark-ignition engines) and very large methyl esters in biodiesel (used as a diesel fuel replacement in diesel engines). The biofuels are oxygenated fuels, which distinguishes them from hydrocarbons in conventional petroleum-based fuels.

The Chemistry of Biodiesel

Differences between Biodiesel, Diesel and Vegetable Oil

A typical molecule of biodiesel looks like the structure below. Mostly it is a long chain of carbon atoms, with hydrogen atoms attached, and at one end is what we call an ester functional group (shown in blue).

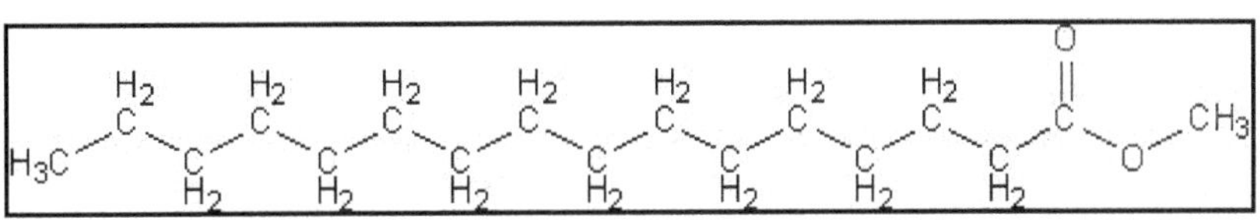

Diesel engines can burn biodiesel fuel with no modifications (except for replacing some rubber tubing that may soften with biodiesel). This is possible because biodiesel is chemically very similar to regular diesel, shown below. Notice that regular diesel also has the long chain of carbon and hydrogen atoms, but doesn't have the ester group shown in blue above.

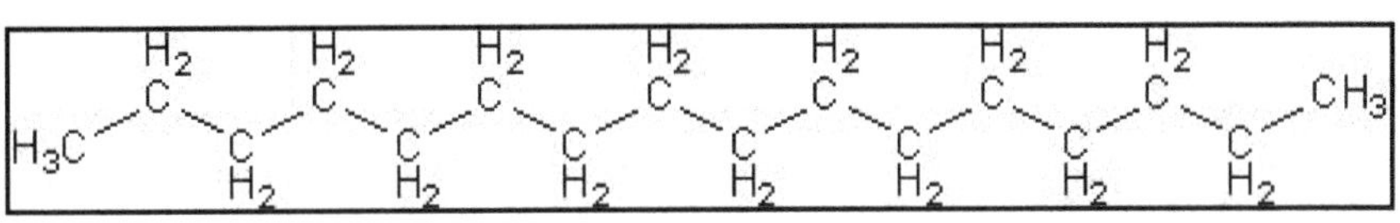

Actually, the first diesel engines didn't run on "diesel" fuel, but on vegetable oil, a sample molecule of which is shown below. Notice that it also has the long rows of carbon and hydrogen atoms, but is about three times larger than normal diesel molecules. It also has ester functional groups (in blue), like biodiesel.

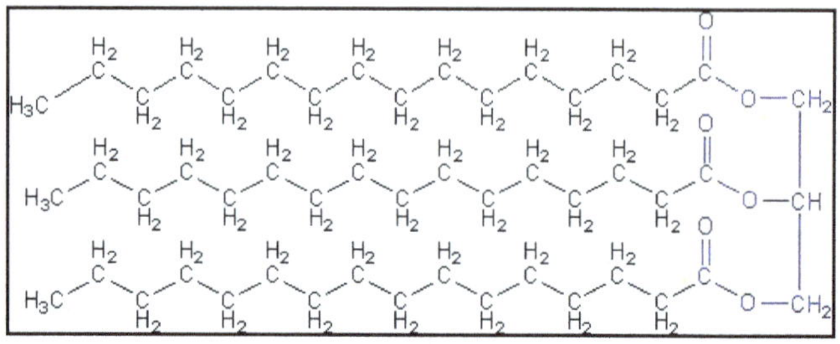

That larger size of vegetable oil means that in cold weather it gels, making it hard to use in an engine. Converting it into biodiesel makes it a smaller molecule, closer to the size of regular diesel, so that it has to get colder than vegetable oil before it starts to gel.

Chemical Conversion of Vegetable Oil to Biodiesel

Vegetable oil, like biodiesel, belongs to a category of compounds called esters. Therefore, converting vegetable oil into biodiesel is called a transesterification reaction. Doing this reaction requires using methanol (shown in green), which causes the red bonds in the structure below to break. This breaks off the blue section, like a backbone on the molecule, which becomes glycerol. The red bonds that did go to the glycerol backbone are placed with bonds to methoxy groups, shown in green in the final structure, that came from the methanol:

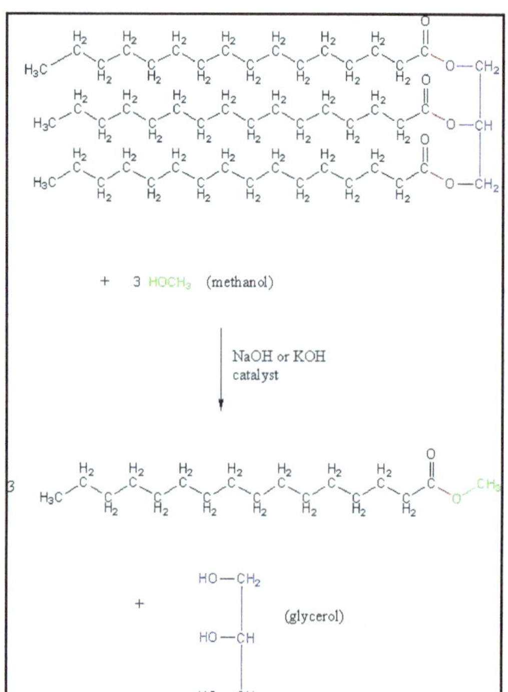

Other Steps in Making Biodiesel

Making good biodiesel requires several other steps besides the transesterification reaction. The first is to remove any traces of water in the vegetable oil. If this is not done, the water will later react with the vegetable oil in the reaction and make soap, shown below.

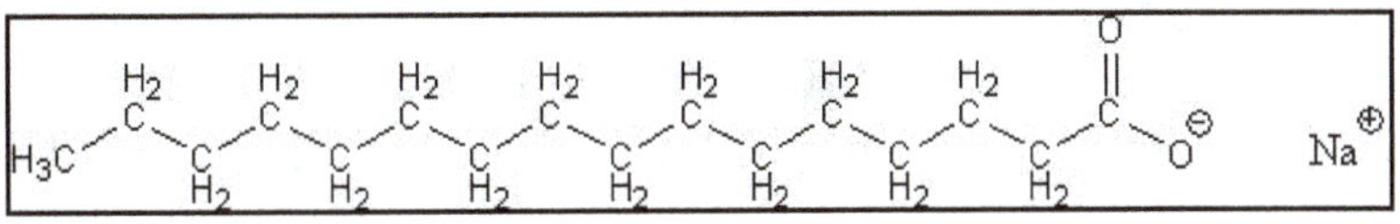

If soap gets made, then later it complicates the steps after the transesterication reaction that are needed to separate the biodiesel from leftover methanol, the NaOH or KOH catalyst, and the glycerol byproduct.

Ethanol, Butanol and Gasoline

The biggest difference between these biofuels (ethanol and butanol) and biodiesel, is that organisms can produce ethanol and butanol on their own. This means there is no need to harvest an organism to obtain a raw product that is converted to biofuel. The advantage of such an approach is that fewer nutrients and less water are required.

The molecular structure of ethanol, butanol, and gasoline are diagramed below. Note that the only real difference, structurally, between ethanol and butanol is the number of carbon atoms they contain. Note that octane is used to represent gasoline even though gasoline is actually a mixture of hydrocarbon molecules ranging in size from four to twelve carbon atoms long.

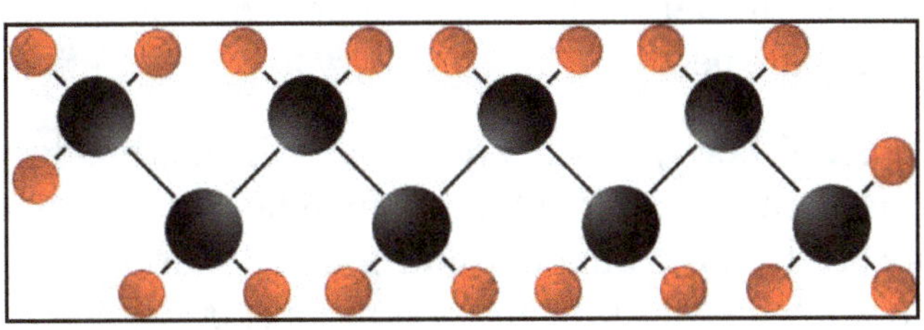

Octane, one of several molecules in gasoline.

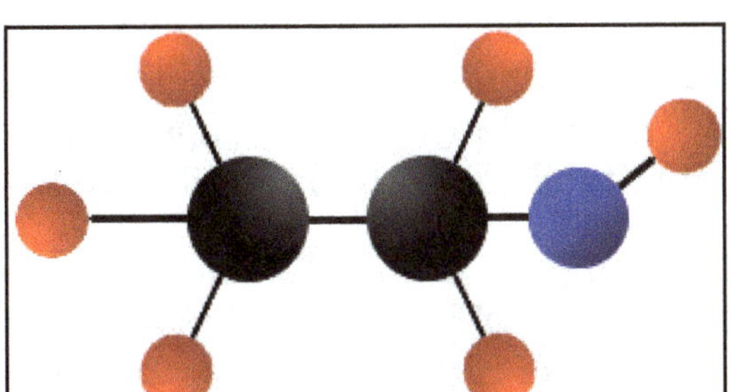

Ethanol (blue represents oxygen).

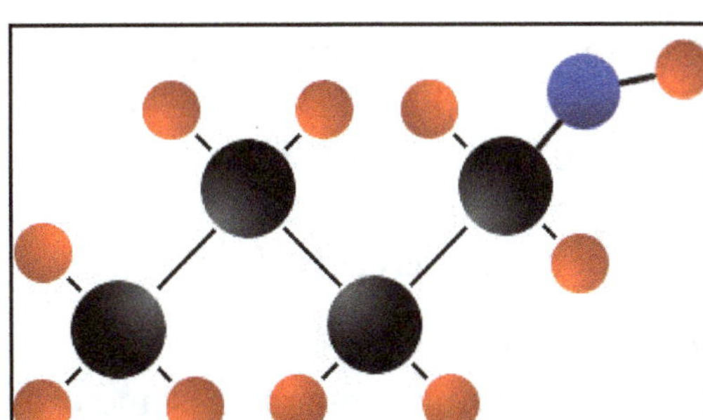

Butanol (blue represents oxygen).

Ethanol and butanol are both considered biofuels because they can be produced from living or recently living organisms. Like the differences between petrodiesel and biodiesel, the big difference between ethanol/butanol and gasoline is the presence of oxygen.

In the case of alcohols (ethanol and butanol are alcohols) there is no ester group, but there is an alcohol group (just an oxygen with a hydrogen attached). This difference means that alcohols behave differently from the esters that make up biodiesel. The chart below compares gasoline to ethanol, butanol, and biodiesel.

Property	Gasoline	Ethanol	Butanol	Biodiesel
Octane (average)	85-96	99.5	97	(cetane) 50-65
Energy Density (MJ/kg)	33	20	30	38

The thing to note is that ethanol has a very low energy density compared to gasoline, but butanol is pretty comparable. It requires about 1.5 times as much ethanol as it does gasoline to generate the same power. In a car, that means you need 1.5 gallons of ethanol to travel the same distance as you would on one gallon of gasoline.

Ethanol versus Butanol

Clearly butanol is more similar to gasoline than is ethanol. The problem is that producing butanol is more difficult and has not been reliably demonstrated on a large scale. Many organisms, like yeast and even some types algae, produce ethanol as part of their normal chemistry. That is to say, a number of organisms produce ethanol just like plants produce oxygen or humans produce carbon dioxide. Many organisms only need carbon dioxide and an energy source to produce ethanol. In the case of algae, only water, sunlight, and $CO_2$are needed. Thus, it is easy to produce ethanol.

The problem with ethanol, beyond energy density, is that it is highly corrosive to engine component, particularly rubbers, and it takes a great deal of energy to produce. Changing feedstock from corn to something like Jatropha or simply using algae can help to offset this energy input, but so far it seems that more energy is needed to produce ethanol than is derived from it. Besides, algae are better at producing biodiesel.

Butanol is less corrosive than ethanol and because it takes about the same amount of energy to produce, it is more efficient in the long run. The problem with butanol production is that it is difficult to do on large scale. Recently, several companies have devised plans to produce butanol in large quantities and are moving forward with plans to make it a commercially viable fuel.

Like biodiesel, both ethanol and butanol are biodegradable. Unlike biodiesel, they do not freeze at lower temperatures and are more stable during long term storage. Thus, these biofuels may offer better alternatives to gasoline, in the long run, than biodiesel offers compared to petrodiesel.

Burning of Biofuels

Burning Alcohol

Ethanol, methanol, butanol, and other alcohols are all flammable. Interestingly, both alcohols and gasoline don't burn in their liquid forms. That is to say, the liquid forms of these molecules are relatively difficult to burn, why is that?

The answer has to do with oxygen (molecular oxygen). Combustion requires oxygen and only gasoline or alcohol exposed to oxygen can burn. In the liquid form, both substances are packed tight enough to prevent too much oxygen from entering the liquid. This is why gasoline burns better when sprayed into an engine by a fuel injector. It is also why the surface of ethanol burns and not the entire liquid contents. So, both fuels need to be aerosolized in order to burn efficiently.

When ethanol is burned in the presence of oxygen, we get an equation that looks something like this.

$$C_2H_5OH + 3O_2 \rightarrow 2CO_2 + 3H_2O$$

Combustion of Ethanol:

When butanol is burned, we see this.

$$2C_4H_9OH + 5O_2 \rightarrow 8CO_2 + 10H_2O$$

Combustion of Butanol:

These equations are, of course, the ideal relationships. Because alcohol contains very little sulfur, there is very little sulfur dioxide produced when it is burned and thus little sulfuric acid. However, alcohol is not a "clean-burning fuel." Alcohol, and any fuel derived from recently living plant and animal matter, contains larger amounts of nitrogen. This means that biofuels produce more nitric oxide and other nitrogen compounds when burned. In some cases, this not only offsets the savings from not producing sulfur compounds, but actually worsens the long term trend for acid rain.

Alcohols are not immune to inefficient combustion issues either. The inefficient combustion of alcohol produces carbon monoxide, formaldehyde, ammonia, benzene, and other toxic chemicals. In some studies, the combustion of ethanol actually produced more formaldehyde, a toxic and carcinogenic compound, than burning gasoline. Ultimately, burning ethanol does produce less carbon monoxide because the fuel itself supplies some oxygen in addition to what is found in the atmosphere. The benefits in terms of carbon monoxide are less with butanol because the ratio of oxygen to carbon is lower in butanol. In ethanol, the ratio is one oxygen molecule to two carbon molecules or 1:2. In butanol, the ratio is 1:4.

We can do a bit of analysis to determine how much byproduct ethanol produces in comparison to gasoline. This comparison will only be for the actual combustion step and does include production, distribution, etc.

When a kilogram of gasoline is burned, we get approximately 33 megajoules of energy and somewhere on the order of 3.09 kilograms of carbon dioxide. When a kilogram of ethanol is burned, we get 20 megajoules of energy and about 1.91 kilograms of carbon dioxide. However, remember that we get 20/33 times the amount of energy from ethanol or only about 0.61 times the amount of energy. This means we need to burn 1.64 kilograms of ethanol to get the equivalent amount of energy, which brings us up to 3.14 kilograms of carbon dioxide. So, ethanol produces MORE carbon dioxide than gasoline.

For butanol, about 2.37 kilograms of carbon dioxide are produce per kilogram burned. Because it returns somewhere on the order of 29 MJ/kilogram, we only need about 1.14 kilograms to get the same energy content as gasoline. Thus, for the same energy content, butanol produces only about 2.7 kg of carbon dioxide, putting it ahead of gasoline and well ahead of ethanol.

Burning Biodiesel

There are two types of biodiesel that are generally used when calculating combustion, the C19 and the C20 chain lengths. We will focus just on C19 (nineteen carbons in the chain) to keep things simple. The equation for ideal combustion of biodiesel looks like this.

$$C_{19}H_{36}O_2 + 27O_2 \rightarrow 19CO_2 + 18H_2O$$

Ideal Combustion of Biodiesel:

If we run the math on this equation, we find that biodiesel produces about 2.52 kilograms of carbon dioxide for every kilogram of fuel burned (2.59 if we use C20). This compares very favorably to petrodiesel, which produces 3.17 kilograms of carbon dioxide per kilogram of fuel. If you figure in the reductions in sulfur emissions, then biodiesel seems to be faring quite well (at least better than ethanol). Of course, we cannot forget to include the energy conversion, which is about 38 MJ/kg for biodiesel and about 43 MJ/kg for petrodiesel. Thus, we need about 1.13 times as much biodiesel to get the same amount of energy and will therefore produce around 2.86 kg of carbon dioxide.

The biggest problem with biodiesel is the production of nitrogen compounds, namely nitric oxide, which are poisonous in and of themselves and contribute to the production of acid rain. In a strange twist of irony, it has been shown that varying the temperature in a diesel engine can reduce the production of nitrogen compounds only to increase the production of soot. Thus, there is a tradeoff that seems to greatly limit the utility of biofuel in reducing overall emissions.

Biofuel Production

Production of Biodiesel

The raw materials for biodiesel production are vegetable oils, animal fats and short chain alcohols. The oils most used for worldwide biodiesel production are rapeseed (mainly in the European Union countries), soybean (Argentina and the United States of America), palm (Asian and Central American countries) and sunflower, although other oils are also used, including peanut, linseed, safflower, used vegetable oils, and also animal fats. Methanol is the most frequently used alcohol although ethanol can also be used.

Since cost is the main concern in biodiesel production and trading (mainly due to oil prices), the use of non-edible vegetable oils has been studied for several years with good results.

Besides its lower cost, another undeniable advantage of non-edible oils for biodiesel production lies in the fact that no foodstuffs are spent to produce fuel. These and other reasons have led to medium- and large-scale biodiesel production trials in several countries, using non-edible oils such as castor oil, tung, cotton, jojoba and jatropha. Animal fats are also an interesting option, especially in countries with plenty of livestock resources, although it is necessary to carry out preliminary treatment since they are solid; furthermore, highly acidic grease from cattle, pork, poultry, and fish can be used.

Microalgae appear to be a very important alternative for future biodiesel production due to their very high oil yield; however, it must be taken into account that only some species are useful for biofuel production.

Although the properties of oils and fats used as raw materials may differ, the properties of biodiesel must be the same, complying with the requirements set by international standards.

Characteristics of Oils and Fats used in Biodiesel Production

Oils and fats, known as lipids, are hydrophobic substances insoluble in water and are of animal or vegetal origin. They differ in their physical states at room temperature. From a chemical viewpoint, lipids are fatty glycerol esters known as triglycerides. The general chemical formula is shown in figure below.

In the figure below, R1, R2 y R3 represent hydrocarbon chains of fatty acids, which in most cases vary in length from 12 to 18 carbon atoms. The three hydrocarbon chains may be of equal or different lengths, depending on the type of oil; they may also differ on the number of double-covalent bonds in each chain. Fatty acids may be saturated fatty acids (SFA) or non-saturated fatty acids (NSFA). In the former, there are only single covalent bonds in the molecules.

$$\begin{array}{l} H_2C-OCOR_1 \\ \quad | \\ HC-OCOR_2 \\ \quad | \\ H_2C-OCR_3 \end{array}$$

General chemical formula of triglycerides.

Table: Chemical formulas of the main fatty acids in vegetable oils.

Fattyacid	Chemical formula
Lauric	(12:0) CH3 (CH2)10 COOH
Palmitic	(16:0) CH3 (CH2)14 COOH
Estearic	(18:0) CH3 (CH2)16 COOH
Oleic	(18:1) CH3 (CH2)7 CH = CH (CH2)7 COOH
Linoleic	(18:2) CH3 (CH2)4 CH = CH CH2 CH = CH (CH2)7 COOH
Linolenic	(18:3) CH3 CH2 (CH = CH CH2)3 (CH2)6 COOH
Erucic	(22:1) CH3 (CH2)7 CH = CH (CH2)11 COOH
Ricinoleic	(18:1) CH3 (CH2)5 CHOH CH2 CH = CH (CH2)7 COOH

Table: Approximate content (in weight) of saturated and non-saturated fatty acids in some vegetable oils and animal fats.

Oil/fat	SFA (& % w/w)	NSFA (& % w/w)
Coconut	90	10
Corn	13	87
Cottonseed	26	74
Olive	14	86
Palm	49	51
Peanut	17	83
Rapeseed	6	94
Soybean	14	86
Sunflower	11	89
Safflower	9	91
Castor	2	98

Yellow grease	33	67
Lard	41	59
Beef tallow	48	52

names of the most important fatty acids in oils are listed in first table along with their chemical formulas. The notation x:y indicates the number of carbon atoms in the oil molecule (x) and the number of unsaturations, i.e. double-covalent bonds (y). For instance, y = 0 for all the SFAs. Table above indicates the approximate content (in weight) of saturated and non-saturated fatty acids in some vegetable oils and animal fats.

The most frequent fatty acids in oils are lauric, palmitic, estearic, linoleic and linolenic, although others may also be present. It is important to note that vegetable oils differ in their content of fatty acids. For instance, ricinoleic acid is the main component of castor oil, whereas in olive oil it is oleic acid, in soybean oil it is linoleic acid, and in linseed oil it is linolenic acid.

The compositions indicated in table above do not discriminate between the different saturated or unsaturated fatty acids. For instance, coconut oil has about 90% of SFAs in its composition (more than half being lauric acid), and palm oil has

$$\begin{array}{ll} H_2C-O-COR & H_2C-O-COR \\ \quad | & \quad | \\ HC-O-COR & CHOH \\ \quad | & \quad | \\ H_2C-OH & H_2C-O-COR \end{array}$$

Chemical formula of diglycerides.

$$\begin{array}{ll} H_2C-O-COR & H_2C-OH \\ \quad | & \quad | \\ CHOH & H_2C-O-COR \\ \quad | & \quad | \\ H_2C-OH & H_2C-OH \end{array}$$

Chemical formula of monoglycerides.

about 49% SFAs (more than 80% palmitic acid). Similarly, 60% of NSFAs content in soybean oil is linoleic acid, while in peanut more than 50% is oleic.

The US Department of Energy indicates that a perfect biodiesel should only comprise mono-unsaturated fatty acids.

Vegetable oils may also contain small percentages of monoglycerides and diglycerides. Their chemical formulae are shown in above two figures. In addition, there will also be small amounts of free fatty acids (in most vegetable oils, less than 1%, except for palm oil, where they can reach up to 15%).

The composition of vegetable oils influences their properties. For instance, the pour point and cloud point temperatures, cetane number and the iodine index depend on the number of unsaturations and the length of the fatty acid chains. A higher content of double-covalent bonds gives a lower solidification point and a higher iodine index.

Characteristics of Alcohols used in Biodiesel Production

Alcohols that can be used in biodiesel production are those with short chains, including methanol, ethanol, butanol, and amylic alcohol. The most widely used alcohols are methanol (CH3OH) and ethanol (C_2H_5OH) because of their low cost and properties. Methanol is often preferred to ethanol in spite of its high toxicity because its use in biodiesel production requires simpler technology; excess alcohol may be recovered at a low cost and higher reaction speeds are reached.

It must be remembered that in order for biodiesel to be a fully renewable fuel, it should be obtained from vegetable oils and animal fats, together with an alcohol that is produced from biomass, such as bioethanol, instead of being a petrochemical product. Several countries are carrying out research towards this objective, such as Spain and Brazil.

Biodiesel Production Process

Biodiesel is produced from vegetable oils or animal fats and an alcohol, through a transesterification reaction. This chemical reaction converts an ester (vegetable oil or animal fat) into a mixture of esters of the fatty acids that makes up the oil (or fat). Biodiesel is obtained from the purification of the mixture of fatty acid methyl esters (FAME). A catalyst is used to accelerate the reaction.

According to the catalyst used, transesterification can be basic, acidic or enzymatic, the former being the most frequently used.

A generic transesterification reaction is presented in Eq.; RCOOR' indicates an ester, R"OH an alcohol, R'OH another alcohol (glycerol), RCOOR" an ester mixture and cat a catalyst:

$$\mathrm{RCOOR'+R''OH,}\ \overset{\Leftrightarrow}{cat}\ \mathrm{R'OH+RCOOR''}$$

When methanol is the alcohol used in the transesterification process, the product of the reaction is a mixture of methyl esters; similarly, if ethanol were used, the reaction product would be a mixture of ethyl esters. In both cases, glycerin will be the co-product of the reaction. This is shown schematically in figures below.

$$\begin{array}{lcccl} H_2C-OCOR_1 & & & H_2C-OH & CH_3-OCOR_1 \\ | & & & | & \\ HC-OCOR_2 & +3\,CH_3OH & \overset{NaOH}{} & HC-OH \quad + & CH_3-OCOR_2 \\ | & & & | & \\ H_2C-OCOR_3 & & & H_2C-OH & CH_3-OCOR_3 \end{array}$$

Basic transesterifi-cation reaction with methanol.

$$\begin{array}{lcccl} H_2C-OCOR_1 & & & H_2C-OH & CH_3CH_2-OCOR_1 \\ | & & & | & \\ HC-OCOR_2 & +3\,CH_3CH_2OH & \overset{NaOH}{} & HC-OH \quad + & CH_3CH_2-OCOR_2 \\ | & & & | & \\ H_2C-OCOR_3 & & & H_2C-OH & CH_3CH_2-OCOR_3 \end{array}$$

Basic transesterification reaction with ethanol.

Although transesterification is the most important step in biodiesel production (since it originates the mixture of esters), additional steps are necessary to obtain a product that complies with international standards. In consequence, once the chemical reaction is completed and the two phases (mix of esters and glycerin) are separated, the mix of methyl esters must be purified to reduce the concentration of contaminants to acceptable levels. These include remnants of catalyst, water and methanol; the latter is usually mixed in excess proportion with the raw materials in order to achieve higher conversion efficiency in the transesterification reaction.

Treatment of Raw Materials

The content of free fatty acids, water and non-saponificable substances are key parameters to achieve high conversion efficiency in the transesterification reaction.

The use of basic catalysts in triglycerides with high content of free fatty acids is not advisable, since part of the latter reacts with the catalyst to form soaps. In consequence, part of the catalyst is spent, and it is no longer available for transesterification. In summary the efficiency of the reaction diminishes with the increase of the acidity of the oil; basic transesterification is viable if the content of free fatty acids (FFAs) is less than 2%. In the case of highly acidic raw materials (animal fats from cattle, poultry, pork; vegetable oils from cotton, coconut, most used oils, etc.) an acid transesterification is necessary as a preliminary stage, to reduce the level of FFAs to the above-mentioned value.

Besides having low humidity and acid content, it is important that the oil presents a low level of non-saponificable substances. If the latter were to be present in significant amounts and soluble in biodiesel, it would reduce the level of esters in the product, making it difficult to comply with the minimum ester content required by the standards.

The AOCS standards list the required properties of oils. Anyway, the properties required by the oils are finally determined by the biodiesel industry in each country. For instance, in Argentina the oils for biodiesel production usually have:

- Acidity level <0.1 mg KOH/g.
- Humidity <500 ppm.
- Peroxide index <10 meq/kg.
- Non-saponificable substances <1%.

Alcohol-catalyst Mixing

The alcohol used for biodiesel production must be mixed with the catalyst before adding the oil. The mixture is stirred until the catalyst is completely dissolved in the alcohol. It must be noted that the alcohol must be water-free (anhydrous) for the reasons explained in the previous paragraph.

Sodium and potassium hydroxides are among the most widely used basic catalysts. For production on an industrial scale, sodium or potassium methoxides or methylates are commercially available.

Of course, due caution must be exercised, and all applicable safety regulations must be followed, when working with methanol, hydroxides and methoxides, independently of the production scale.

The alcohol-to-oil volume ratio, R, is another key variable of the transesterification process. The stoichiometric ratio requires 1 mol of oil to react with 3 mol of alcohol, to obtain 3 mol of fatty acids methyl esters (FAME) and 1 mol of glycerin. However, since the reaction is reversible, excess alcohol as a reactant will shift the equilibrium to the right side of the equation, increasing the amount of products (as it may be inferred from Le Chatelier's principle). Although a high alcohol-to-oil ratio does not alter the properties of FAME, it will make the separation of biodiesel from glycerin more difficult, since it will increase the solubility of the former in the latter. Usually, a 100% alcohol excess is used in practice, that is, 6 mol of alcohol per mole of oil. This corresponds to a 1:4 alcohol-to-oil volume ratio (R = 0.25). Finally, it must be noted that the necessary amount of catalyst is determined taking into account the acidity of the oil, by titration.

Chemical Reaction

The chemical reaction takes place when the oil is mixed with the alkoxide (alcohol–catalyst mix) described in the previous paragraph. This requires certain conditions of time, temperature and stirring. Since alcohols and oils do not mix at room temperature, the chemical reaction is usually carried out at a higher temperature and under continuous stirring, to increase the mass transfer between the phases.

Usually, emulsions form during the course of the reaction; these are much easier and quicker to destabilize when methanol is used, in comparison to ethanol. Due to the greater stability of emulsions formed, difficulties arise in the phase separation and purification of biodiesel when ethanol is used in the reaction.

The transesterification process may be carried out at different temperatures. For the same reaction time, the conversion is greater at higher temperatures. Since the boiling point of methanol is approximately 68C (341 K), the temperature for transesterification at atmospheric pressure is usually in the range between 50 and 60C.

It is very useful to know the chemical composition of the mixture during the reaction; then, if the reaction mechanism and kinetics are known, the process can be optimized. However, the determination of the mixture composition is not easy, since more than a hundred substances are known to be present. For instance, for biodiesel production from rapeseed oil (whose main SFAs are palmitic, oleic, linoleic and linolenic) and methanol, with potassium hydroxide as a catalyst, it could be theoretically possible to find 64 isomers of triglycerides, 32 diglycerides, 8 monoglycerides, their methyl esters, potassium salts of the fatty acids, potassium methoxide, water, etc.

The studies on this subject indicate the following general guidelines:

$$TG + MOH \xrightarrow{KOH} KOH\ DG + ME$$

$$DG + MOH \xrightarrow{KOH} KOH\ MG + ME$$

$$MG + MOH \xrightarrow{KOH} KOH\ G + ME$$

where MOH indicates methanol, ME are the methyl esters, TG, DG and MG are tri-, di- and monoglycerides, respectively, and G is the glycerin.

Several methods, with different levels of equipment complexity and training requirements, have been devised to analyze samples that are mixtures of fatty acids and esters from mono-, di-, and triglycerides obtained from transesterification of vegetable oils.

It must be noted that thin layer chromatography (TLC) provides essentially qualitative information about the sample composition, as distinct from the other methods, that can be used for quantitative analysis. However, the simplicity, speed and low cost of TLC make it quite attractive as a technique for process optimization and routine checks, especially in small- and medium-scale production plants, and also for training purposes.

Catalysts

The catalysts used for the transesterification of triglycerides may be classified as basic, acid or enzymatic.

Basic catalysts include sodium hydroxide (NaOH), potassium hydroxide (KOH), carbonates and their corresponding alcoxides (for instance, sodium methoxide or ethoxide). There are many references on basic catalysts in the scientific literature.

Acid catalysts include sulfuric acid, sulfonic acids and hydrochloric acid; their use has been less studied.

Heterogeneous catalysts that have been considered for biodiesel production include enzymes, titanium silicates, and compounds from alkaline earth metals, anion exchange resins and guanidines in organic polymers. Lipases are the most frequently used enzymes for biodiesel production.

Separation of the Reaction Products

The separation of reaction products takes place by decantation: the mixture of fatty acids methyl esters (FAME) separates from glycerin forming two phases, since they have different densities; the two phases begin to form immediately after the stirring of the mixture is stopped. Due to their different chemical affinities, most of the catalyst and excess alcohol will concentrate in the lower phase (glycerin), while most of the mono-, di-, and triglycerides will concentrate in the upper phase (FAME). Once the interphase is clearly and completely defined, the two phases may be physically separated. It must be noted that if decantation takes place due to the action of gravity alone, it will take several hours to complete. This constitutes a "bottleneck" in the production process, and in consequence the exit stream from the transesterification reactor is split into several containers. Centrifugation is a faster, albeit more expensive alternative.

After the separation of glycerin, the FAME mixture contains impurities such as remnants of alcohol, catalyst and mono-, di-, and triglycerides. These impurities confer undesirable characteristics to FAME, for instance, increased cloud point and pour point, lower flash point, etc. In consequence a purification process is necessary for the final product to comply with standards.

Purification of the Reaction Products

The mixture of fatty acids methyl esters (FAME) obtained from the transesterification reaction must be purified in order to comply with established quality standards for biodiesel. Therefore, FAME must be washed, neutralized and dried.

Successive washing steps with water remove the remains of methanol, catalyst and glycerin, since these contaminants are water-soluble. Care must be taken to avoid the formation of emulsions during the washing steps, since they would reduce the efficiency of the process. The first washing step is carried out with acidified water, to neutralize the mixture of esters. Then, two additional washing steps are made with water only. Finally the traces of water must be eliminated by a drying step. After drying, the purified product is ready for characterization as biodiesel according to international standards.

An alternative to the purification process described above is the use of ion exchange resins or silicates.

Glycerin as obtained from the chemical reaction is not of high quality and has no commercial value. Therefore, it must be purified after the phase separation. This is not economically viable in small scale production, due to the small glycerine yield. However, purification is a very interesting alternative for large-scale production plants, since, in addition to the high quality glycerin, part of the methanol is recovered for reutilization in the transesterification reaction (both from FAME and glycerin), and thus lowering biodiesel production costs. The steady increase of biodiesel production is fostering research for novel uses of glycerin in the production of high-value-added products.

It must be noted that the stages of the biodiesel production process are the same for all the production scales (laboratory, pilot plant small-, medium-, and large-scale industrial). However, the necessary equipment will be significantly different.

Production of Bioethanol

Ethanol can be produced from biomass by the hydrolysis and sugar fermentation processes. Biomass wastes contain a complex mixture of carbohydrate polymers from the plant cell walls known as cellulose, hemi cellulose and lignin. In order to produce sugars from the biomass, the biomass is pre-treated with acids or enzymes in order to reduce the size of the feedstock and to open up the plant structure. The cellulose and the hemi cellulose portions are broken down (hydrolysed) by enzymes or dilute acids into sucrose sugar that is then fermented into ethanol. The lignin which is also present in the biomass is normally used as a fuel for the ethanol production plants boilers. There are three principle methods of extracting sugars from biomass. These are concentrated acid hydrolysis, dilute acid hydrolysis and enzymatic hydrolysis.

Concentrated Acid Hydrolysis Process

The Arkanol process works by adding 70-77% sulphuric acid to the biomass that has been dried to a 10% moisture content. The acid is added in the ratio of 1.25 acid to 1 biomass and the temperature is controlled to 50C. Water is then added to dilute the acid to 20-30% and the mixture is again heated to 100C for 1 hour. The gel produced from this mixture is then pressed to release an acid sugar mixture and a chromatographic column is used to separate the acid and sugar mixture.

Dilute Acid Hydrolysis

The dilute acid hydrolysis process is one of the oldest, simplest and most efficient methods of producing ethanol from biomass. Dilute acid is used to hydrolyse the biomass to sucrose. The first stage uses 0.7% sulphuric acid at 190C to hydrolyse the hemi cellulose present in the biomass. The second stage is optimised to yield the more resistant cellulose fraction. This is achieved by using 0.4% sulphuric acid at 215C.The liquid hydrolates are then neutralised and recovered from the process.

Enzymatic Hydrolysis

Instead of using acid to hydrolyse the biomass into sucrose, we can use enzymes to break down the biomass in a similar way. However this process is very expensive and is still in its early stages of development.

Wet Milling Processes

Corn can be processed into ethanol by either the dry milling or the wet milling process. In the wet milling process, the corn kernel is steeped in warm water, this helps to break down the proteins and release the starch present in the corn and helps to soften the kernel for the milling process. The corn is then milled to produce germ, fibre and starch products. The germ is extracted to produce corn oil and the starch fraction undergoes centrifugation and saccharifcation to produce gluten wet cake. The ethanol is then extracted by the distillation process. The wet milling process is normally used in factories producing several hundred million gallons of ethanol every Year.

Dry Milling Process

The dry milling process involves cleaning and breaking down the corn kernel into fine particles using a hammer mill process. This creates a powder with a course flour type consistency. The powder contains the corn germ, starch and fibre. In order to produce a sugar solution the mixture is then hydrolysed or broken down into sucrose sugars using enzymes or a dilute acid. The mixture is then cooled and yeast is added in order to ferment the mixture into ethanol. The dry milling process is normally used in factories producing less than 50 million gallons of ethanol every Year.

Sugar Fermentation Process

The hydrolysis process breaks down the cellulostic part of the biomass or corn into sugar solutions that can then be fermented into ethanol. Yeast is added to the solution, which is then heated. The yeast contains an enzyme called invertase, which acts as a catalyst and helps to convert the sucrose sugars into glucose and fructose (both $C6H12O6$).

The chemical reaction is shown below:

			Invertase			
$C_{12}H_{22}O_{11}$	+	H_2O	→	$C_6H_{12}O_6$	+	$C_6H_{12}O_6$
Sucrose		Water	Catalyst	Fructose		Glucose

The fructose and glucose sugars then react with another enzyme called zymase, which is also contained in the yeast to produce ethanol and carbon dioxide.

The chemical reaction is shown below:

	Zymase			
$C_6H_{12}O_6$	→	$2C_2H_5OH$	+	$2CO_2$
Fructose / Glucose	Catalyst	Ethanol		

The fermentation process takes around three days to complete and is carried out at a temperature of between 250C and 300C.

Fractional Distillation Process

The ethanol, which is produced from the fermentation process, still contains a significant quantity of water, which must be removed. This is achieved by using the fractional distillation process. The distillation process works by boiling the water and ethanol mixture. Since ethanol has a lower boiling point (78.3C) compared to that of water (100C), the ethanol turns into the vapour state before the water and can be condensed and separated.

Bioethanol Production by Algae

Algae as Bioethanol Feedstock

Algae as bioethanol feedstock Algae are simple organisms containing chlorophyll and they use light for photosynthesis. Algae can grow phototrophically or heterotrophically. Phototrophic algae convert carbondioxide in atmosphere to nutrients such as carbohydrate. Conversely, heterotrophic algae continue their development by utilizing organic carbon sources. Algae can grow in every season and everywhere such as salty waters, fresh waters, lakes, deserts and marginal fields etc. However for their cultivation, generally open systems like ponds and photobioreactors as closed systems are used. Open ponds are the most used cultivation systems in industry. They are more preferable than other systems due to having low investment and operation costs. On the other hand difficult control of cultivation conditions and contamination risk are the main disadvantages of the open systems. Besides being cheap and low energy need, their cleaning also can be done easily. Although, open tanks have low cost and easy operation, parameters like light intensity, temperature, pH and dissolved oxygen concentration cannot be controlled easily. Most produced algae species in open systems are Spirulina, Chlorella and Dunaliella. In comparison with open systems, photobioreactors have very high photosynthetic efficiency. Thus, photobioreactors ensures high biomass yield. Though they are expensive, they are preferred for specific algal production. Algal production which is controlled in terms of parameters like light, pH, carbon dioxide etc., can be achieved and also contamination risk is not seen mostly in photobioreactors. Since they are closed systems, evaporation doesn't occurred and they enable production of special biochemical materials. Although there are many types of photobioreactors, most commonly systems are vertical and horizontal tubular columns and flat-type photobioreactors. These photobioreactors which are made of glass or plastic, can be designed in shapes of horizontal vertical, conical or curved etc.

Algae are classified as microalgae and macroalgae. Microalgae as their name implies, are prokaryotic or eukaryotic photosynthetic microorganisms. They can survive in hard conditions with their unicellular or simple colony structures. Because of being photosynthetic organism, they can produce high amount of lipid, protein and carbohydrate in a short time. Besides biodiesel and bioethanol there are lots of high value products and sub-products produced from microalgae such as biogas, biobutanol, acetone, Omega 3 oil, eicosapentaenoic acid, livestock feed, pharmaceuticals and cosmetics. Especially sub-products are preferred for economic support of main process. Chemical composition of microalgae can change according to the cultivation type and cultivation conditions. They can have rich or balanced composition of protein, lipid and carbohydrate amounts. Microalgae especially get attention due to have high lipid content. Many species of microalgae accumulate a significant amount of lipids in their structure and can provide high oil yield. Their average lipid content can change between 1-70%, but this ratio can reach up to 90% of dry weight under certain conditions. Macroalgae or seaweed are plants which are adapted to the marine life, often located in coastal areas. They are classified as brown seaweeds, red seaweeds and green seaweeds according to their pigments. Due to have high photosynthesis capability, they have sufficient carbon source for usage in biorefinery. On the contrary of their appearances, their features of morphologic and physiological and chemical compositions are different from terrestrial plants. Unlike the structure of the lignocellulosic biomass of microalgae, they comprise substances such as carrageenan, laminaran, mannitol, alginate which are used in various sectors. They are separated from microalgae with having low lipid content and different from lignocellulosic material with having less or no lignin in their structure.

Microalgae stand out as biodiesel feedstock with the ability of lipid production and high photosynthetic efficiency. As for macroalgae, they are utilized for biogas or bioethanol production with their carbohydrates. First studies as algal biofuels are focused on biodiesel production. However, there is a potential for carbohydrates in the structure of algae which can be utilized for ethanol production after various hydrolysis processes. Algal cells in the water don't need structural biopolymers such as hemicellulose and lignin which are necessary for terrestrial plants. This simplifies the process of bioethanol production. Marine algae can produce high amount of carbohydrate every year. Also it is expected that algae will meet the demand of biofuel feedstock due to harvest in a short time than other biofuel raw materials. Microalgae which have high amount of starch such as Chlorella, Dunaliella, Chlamydomonas, Scenedesmus are very useful for bioethanol production. In addition to that, microalgae don't need energy consumption for distribution and transportation of molecules like starch. Like microalgae, macroalgae are also raw materials that can be used in ethanol production. Absence of lignin or having less lignin in the structure, simplifies the hydrolysis stages. Although it changes with the algal species, they have various amounts of heteropolysaccharides in their structures. Whereas red algae contain carrageenan and agar, brown algae have laminaran and mannitol in their structure.

Algal Polysaccharides

Ethanol production from algae is based on fermentation of algal polysaccharides which are starch, sugar and cellulose. For microalgae, their carbohydrate content (mostly starch) can be reached to 70% under specific conditions. Microalgal cell walls are divided into inner cell wall layer and outer cell wall layer. Outer cell layer can be trilaminar outer layer and thin outer monolayer. Also there can be no outer layers as well. Outer cell walls of microalgae contain certain polysaccharides such as pectin, agar and alginate. However their composition can be vary from species to species. On the contrary, inner cell walls of microalgae constitute mostly cellulose, hemicellulose and other materials. Due to have cellulose in their cell walls and starch, microalgae are considered as a feedstock for production of bioethanol. Most of their cell wall polysaccharides and starch can be fermented for bioethanol production.

Similarly, carbohydrate content of macroalgae is found 25-50% in the green algae, 30-60% in the red algae and 30-50% in the brown algae. Macroalgae species which have the highest polysaccharide content are Ascophyllum (42-70%), Porphyra (40-76%) and Palmaria (38-74%). High carbohydrate content of algal species are presented in table below. Polysaccharides in the cell wall of macroalgae are composed of cellulose and hemicelluloses. Cellulose and hemicelluloses content of macroalgae compose 2-10% of dry weight. Lignin is only exists in Ulva species and it constitutes 3% of dry weight. Differently from microalgae, alginate, mannitol, glucan and laminarin are the most abundant polymers in macroalgal structure. Alginates are.

Table: Carbohydrate content of algal species.

Algal species	Carbohydrate content (%)
C. vulgaris	55.0
Chlamydomonas reinhardtii	UTEX 90 60.0
Chlorococum sp.	32.5
S. obliquus CNW-N	51.8

Tetraselmis sp.CS-362	26.0
Ulva lactuca	55-60
Ascophyllum	42-70
Porphyra	40-76
Palmaria	38-74

polymers which are obtained from cell walls of various brown algae. They consist of mannuronic acid and L-gluronic acid monomers and they are extracted from cell walls by using sodium carbonate. Although they are usually used as stabilizer in pharmaceutical industry, they also used in paper and adhesive manufacture, oil, photography and textile industries. Caragenan is another polysaccharide which is obtained from red algae. It is used as stabilizer in food, textile and pharmaceutical industry. Agar is also acquired from red algae. Like caragenans and alginates, it is extracted with hot water and used as stabilizer and gelling agent. 90% of produced agar is utilized in the food industry, the remaining 10% is used in microbiological and biotechnological field. Mannitol which is a structure in brown algae is a sugar alcohol, especially found in Laminaria and Ecklonia. Mannitol content of macroalgae can change with season and environmental conditions. Mannitol can be used in pharmaceutical, paint, leather and paper manufacture. In addition to that, mannitol can be utilized in food industry as a coating material. Laminarin is a polysaccharide which helps the immune system by increasing the B cells, provides protection against infection by bacterial pathogens and severe irradiation. Another polysaccharide from macroalgae is ulvan. It is mainly presented in Ulva sp. and it is source of sugars for production of fine chemicals.

Pre-treatment Technologies for Biomass

The most important difficulties encountered in the production of bioethanol are the pretreatment of biomass. The objectives of an effective pre-treatment are obtaining sugars directly or later by hydrolysis, preventing lost or degradation of obtained sugars, limiting the toxic materials which inhibit the ethanol production, reducing energy requirement for process and minimizing the production cost. There are four pre-treatment techniques including physical, chemical, physicochemical and biological pre-treatments that are applied to biomass. Pre-treatment process is the step that forms the significant part of the cost of ethanol production. Although there is no technique that can be considered as the best option, researches and developments are carried on to reduce cost and improve performance.

Physical Pre-treatments

Mechanical Comminution

Chipping, grinding and milling are the most used techniques for mechanical comminution.

Comminutions improve the efficiency of the process for the next steps by reducing the polymerization degree and increase the specific surface by reducing cellulose cristallinity.

Energy that is need in the process depends on the initial and final dimensions of particles, moisture content and structure of the raw material. In order to assist enzymatic hydrolysis of lignocellulosic materials various milling techniques can be used. For instance, pretreatment of rice straw with wet disk milling gave higher hydrolysis yields than usual dry milling.

Pyrolysis

Pyrolysis is an endothermic process which is a reaction needs low energy input and treats biomass over the temperature of 300°C and degrades cellulose to char and gaseous products like CO and H_2. When the char is washed with water or diluted acid, remaining solution contains sufficient amount of carbon source to support microbial growth for the production of bioethanol. Approximately 55% of biomass weight is lost in the washing step. It is reported in a study that Fan et al. have performed 80-85% conversion of cellulose to reducing sugars.

Microwave Oven Pre-treatment

Microwave oven pre-treatment is a simple method with short reaction time, high heating efficiency and low energy input. Thermal and non-thermal effects which are generated by microwaves in a liquid medium are used in this technique. The heat generated in biomass results in a polar bond vibration. This causes an explosion between the particles and degradation of lignocellulosic structure. Asetic acid is released from lignocellulosic material and an acidic medium is occurred for hydrolysis. Ooshima et al. investigated the effect of microwave pre-treatment on rice straw and baggase and it was found that an improvement in total reducing sugar production. In recent years, microwave pre-treatments are carried out with various chemical reagents and their potential are investigated. In the studies of alkali microwave pre-treatment, NaOH provides higher reducing sugar yields on switchgrass and coastal bermudagrass in comparison with other alkaline reagents such as Na_2CO_3, $Ca(OH)_2$. Also for pre-treatment of rice straw and its hulls, this technique made cellulose more accessible to enzymes.

Physicochemical pre-treatments

Steam Explosion Method

Steam explosion method is a technique that provides accessibility on the biomass for degradation of cellulose. This method comprise the heating of biomass under high pressure steam (20–50 bar, 160-270 °C) for a few minutes, then reaction is stopped when the pressure conditions arrive to the atmospheric conditions. Diffusion of the steam into the lignocellulosic matrix leads to the dispersion of fibers. No catalyst is used during the applied method. Levulinic acid, xylitol and alcohols are obtained after the degradation of biomass. Many types of biomass such as poplar, eucalyptus, olive residues, corn stover, wheat straw, sugarcane bagasse, grasses have been pre-treated with steam explosion method efficiently.

Liquid Hot Water Method

Liquid hot water method treats biomassby using water which is kept in a liquid state under high pressure and temperature for 15 minutes without adding any chemical or catalyst. Instead of steam explosion method, this technique does not need rapid pressure drop or expansion. Pressure is used to prevent evaporation and to stabilize the water in this method. Although it provides the release of hemisellulosic sugars as oligomers, it causes the formation of little amounts of undesirable components which inhibit microbial growth such as carboxylic acid, furfural. Since there is no need for chemicals, it is an environmental and economic method. It is reported that liquid hot water method improves the enzymatic hydrolysis by removing 80% of hemicelluloses when it is pre-treated corn stover, sugarcane bagasse and wheat straw.

Ammonia Fiber Explosion (AFEX)

Ammonia fiber explosion (AFEX) is a method that liquid ammonia and steam explosion are carried out together. In this method, biomass which has 15-30% moisture content is treated with liquid ammonia at a loading ratio of 1–2 kg NH3/kg dry biomass. To acquire appropriate temperature, pressure over 12 atm is required. Whereas being an easy method and have short reaction time, it is not effective on raw materials that contain high lignin content. Ammonia has effects such as shredding biomass fibers, partially decrystallization of cellulose and destroying carbohydrate attachments. Although sugars are not released directly with this method, it enhances polymers (hemicellulose and cellulose) to be attacked enzymatically. Thus, low amount of enzyme is enough for enzymatic hydrolysis after AFEX. In order to improve the process economically, ammonia must be recover after the pre-treatment. Ammonia loading, temperature, high pressure, moisture content of biomass, and residence time are the basic parameters which effect AFEX process. Up to 90% cellulose and hemicelluloses conversions can be acquired with this technique.

CO_2 Explosion

CO_2 explosion is similar to AFEX method. However this method has low process cost due to need low temperature. Also formation of inhibitors in the steam explosion is not occurred in this technique. In addition to that, its conversion yields are very high compared to steam explosion.

Wet Oxidation

Wet oxidation method is based on the treatment of biomass with water and air or oxygen as a catalyst over the temperature of 120 °C. Although solubility of hemicellulose and lignin are increased with this method, free hemicelluloses molecules do not hydrolyze. Whereas sugar monomers are formed in steam explosion and dilute acid pre-treatment, sugar which released in wet oxidation method are oligomers. In a study performed by Pederson et al.40% glucose yield was obtained for wet oxidation of wheat straw.

Chemical Pre-treatments

Chemical pre-treatments include dilute acid, alkaline, ammonia, organic solvent pre treatments and methods that use other chemicals. These processes are easy to perform and also good conversion yields are achieved in a short time.

Acid Pre-treatment

Acid pre-treatments are methods that acid is used as catalyst to make cellulose more accessible to the enzymes. These processes are divided into two groups as using concentrated acid or diluted acid. Using concentrated acid is less preferable than dilute acid because of forming high amount of inhibiting components and causing corrosion in the equipments. Generally sulphuric acid, hydrochloric acid, nitric acid and phosphoric acid are used in these pretreatments. Dilute acid are applied at moderate temperatures to convert lignocellulosic structures to soluble sugars. Nowadays biomass is pre-treated with dilute sulphuric mostly to hydrolyze hemicelluloses and facilitate enzymatic hydrolysis. Dilute sulphuric acid hydrolyzes biomass to hemicelluloses, and then hydrolyzes to xylose and other sugar and break xylose down to furfural. Furfural which is a toxic component in ethanol production process, is recovered by distillation. Miranda et al. have investigated the effect of acid pretreatments with the concentrations between 0.05-10 N, and have obtained the highest sugar yield under the condition of 2 N acid pre-treatment. In their experiments, 2 N to 10 N acid pretreatments, it is reported that a decrease have been observed in sugar yields. Larsson et al. also mentioned that in an experiment about acid pre-treatment of soft wood, a decrease in ethanol yields have been observed with an increasing acid concentration. In addition to this, it is indicated that formic acid which is a toxic molecule, is presented in the media and inhibits the fermentation.

Alkaline Pre-treatment

These processes are carried out at low temperature and pressure compared to other techniques. Unlike acid pre-treatments, lignin can be removed without major effects on the other components. However there are limitations such as transformation of some alkaline to unrecoverable salts. In addition to that, solubility of hemicelluloses and cellulose are less in this pre-treatment compared to solubility in acid pre-treatment. Alkaline pre-treatment reduces the lignin and hemicelluloses content of biomass and improves the surface area and helps water molecules for breaking bonds between hemicelluloses and lignin. The most used catalysts in this method are sodium hydroxide, potassium hydroxide, calcium hydroxide and ammonia. Effects of alkaline pre-treatments are varies according to biomass. In an alkaline pretreatment of coastal bermudagrass, reducing sugar yields are decrease with an increasing alkaline concentration. However, Wang et al. reported that under the conditions of increasing alkaline concentrations, glucose yields were increased. Like dilute acid pretreatments, dilute alkaline pre-treatments also can form inhibitory by-products such as furfural, hydroxymethylfurfural and formic acid.

Organosolv Pre-treatment

Organosolv pre-treatment is a process that uses organic solvents such as methanol, ethanol, acetone, ethylene glycol. Catalysts are also can be added to the process along with solvents.

Hydrochloric acid, sulphuric acid, sodium hydroxide and ammonia are the catalysts used in the process. Besides bonds of lignin and hemicellulose can be broken, pure and high quality lignin can be obtained as a by-product. Removal of lignin improves the surface area and provides accessibility of enzymes to cellulose. After the pre-treatment, cellulosic fibers, solid lignin and liquid solution of hemicellulose sugars are obtained. This method has some disadvantages like oxidation, volatilization and creating high risk in process at high pressure.

Also solvents must be recovered due to formation of significant amounts of furfural and soluble phenols and to reduce operation cost.

Biological Pre-treatments

Compared to the above methods applied to the production of bioethanol, using fungi in pretreatments is considered environmentally friendly because of not using chemicals, less energy input, not required reactors that resistant to corrosion and pressure, and minimum inhibitor formation. Fungi which are used in biological pre-treatments are generally brown, white and soft mold. These fungi can be degrade lignin, hemicelluloses and cellulose partially. Despite of its advantages, long process time, large production are and need of control continuously for growth of microorganisms ensue as disadvantages for commercial productions.

Enzymatic hydrolysis is the step of hydrolysis of cellulose by specific cellulase enzymes. Obtained products after hydrolysis are reducing sugars that include glucose. Cost of the enzymatic hydrolysis are less than acid or alkaline hydrolysis due to reaction is carried out under mild conditions (4.8 pH, temperature of 45-50 °C). Cellulase enzymes that are used in hydrolysis can be produced by bacteria and fungi. These microorganisms can be aerobic, anaerobic, mesofilic or thermophilic. Bacteria which produce cellulase can be exemplify as Clostridium, Cellulomonas, Bacillus, Thermomonospora, Ruminococcus, Bacteriodes, Erwinia, Acetovibrio, Microbispora and Streptomyces. Trichoderma, Penicillium, Fusarium, Phanerochaete, Humicola and Schizophillum sp. are identified as cellulase produced fungi among the fungi. Although there are anaerobic bacteria which produce cellulase with high specific activity, these bacteria are not suitable for commercial productions. Cellulase enzymes consist of mixture of endoglucanase, exoglucanase and b-glucosidase. While endoglucanase attacks the regions where cellulose fibers have low crystallinity, exoglucanase removes the cellulose units from released chains with the effect of endoglucanase and then degrades the molecule. B-glucosi-dase hydrolyzes the cellulose units and enables the formation of glucose. Enzymatic hydrolysis can be affected by certain factors which are enzyme-related and substrate-related factors. Substrate-related factors have a directly influence on enzymatic hydrolysis. These factors are connected to each other and effect the enzymatic conversion. These factors can be defined as degree of polymerization and crystallinity of cellulose, accessibility of the substrate, lignin and hemicelluloses content and pore size.

Hydrolysis rates of biomass depend on the degree of polymerization and crystallinity of cellulose. Degree of polymerization is related to crystallinity. Cellulase enzymes can hydrolyze the crystalline structure of cellulose. Endoglucanase enzymes decrease polymerization degree of cellulosic component by cutting the internal sites of cellulose chains in the enzymatic hydrolysis. Accessibility of the substrate is another main factor effect hydrolysis rate. The rate of hydrolysis increases with increasing substrate accessibility because of being surface area more available for enzymatic attack. Lignin and hemicellulose are complex structures to hydrolyze in lignocellulosic materials. Due to have a role like cement, lignin acts as physical barrier and prevents the digestible parts of cellulose to hydrolyze and it becomes very difficult for enzymes to access cellulose. For this reason, they reduce the efficiency of hydrolysis. Removal of hemicellulose enhances the pore size and provides accessibility to cellulose for enzymes in order to perform hydrolysis efficiently. Pore size of the substrate is one of the limiting factors in enzymatic hydrolysis process. In many lignocellulosic material, external area of the biomass is smaller than internal area and this situation causes cellulase enzymes to entrap in the pores of the material. In order to increase hydrolysis rate, porosity of the biomass should be increased.

Fermentation

Fermentation is a process that based on disciplines of chemistry, biochemistry and microbiology and which fermentable sugars are converted to ethanol by microorganisms. Process consists of conversion of glucose to alcohol and carbon dioxide:

$$C_6H_{12}O_6 \rightarrow 2C_2H_5OH + 2CO_2$$

In this process 0.51 kg bioethanol and 0.49 kg carbon dioxide are obtained from per kg of glucose in theory maximum yield. However practically, microorganisms also use glucose for their growth, the actual yield is less than 100%. Microorganisms used in fermentation are utilized from 6-carbon sugars in ethanol production. Therefore, cellulosic biomass which have high amount of glucose are the materials that have easiest conversion capability. One of the most effective yeast which produces bioethanol is Saccharomyces cerevisiae. Besides having high bioethanol production yields, it has a resistance to high bioethanol concentration and inhibitor components which can be occurred after acid hydrolization of lignocellulosic biomass. Because reaction occurs under anaerobic conditions, oxygen molecules must be removed with nitrogen gas as a swept gas. Yeast and fungi can tolerate 3.5-5.0 pH ranges. S.cerevisiae has high osmotic resistance and can tolerate low pH levels like 4.0. Zymomonas stands out with rapid bioethanol production and high productivity compared to other traditional yeasts. However Z.mobilis cannot tolerate the toxic effects of asetic acid and various phenolic compounds in the lignocellulosic hydrolysate. Bioethanol yields of microorganisms are depend on temperature, pH level, alcohol tolerance, osmotic tolerance, resistance for inhibitors, growth rate and genetic stability. Fermentation processes generally are carried out with two basic processes as simultaneous saccharification and fermentation and separate hydrolysis and fermentation, yet new production processes have been developed.

Separate Hydrolysis and Fermentation (SHF)

Enzymatic hydrolysis is performed separately from fermentation in this process. Liquid which comes from hydrolysis reactor first converted to ethanol in a reactor that glucose fermented in, and then ethanol is distilled and remained unconverted ksilose is converted to ethanol in a second reactor. Advantage of the process is performing reactions in optimum conditions. On the other hand, usage of different reactors is increasing the cost. Also glucose and cellulose units that obtained after hydrolysis, inhibit activity of the enzyme and decrease hydrolysis rate.

Simultaneous Saccharification and Fermentation (SSF)

In this process, pre-treatment and enzymatic hydrolysis steps are carried out with fermentation step in the same reactor. It is very efficient when dilute acid or hot water at high temperature is applied in the process. High bioethanol yields can be achieved with SSF process. Also inhibiton of enzyme activity is very low due to fermenting glucose and cellulose units in the same media by yeast. Therefore, this process needs low amount of enzyme. In addition to that, process cost is reduced because of the reactions are carried out in one reactor. As a disadvantage, temperatures differences between saccharification and fermentation cause various effects in growth of microorganisms. Saccharomyces cultures are used in pH of 4.5 and temperature of 37 °C this process.

Simultaneous Saccharification and Co-Fermentation (SSCF) & Separate Hydrolysis and Co-Fermentation (SHCF).

Saccharomyces cerevisiae which used in fermentation cannot convert carbohydrates like pentos under moderate conditions and this causes impurity for biomass and decreases bioethanol yield. In order to overcome this, recombinant yeasts can be used to convert residues such as pentose to ethanol. In SSCF, recombinant yeasts and cellulase enzyme complex are fed to the same vessel to convert biomass to ethanol. This system is generally the same as SSF process. SCHF process is a combination of SSCF and SHF. In this process, fermentation and hydrolysis are carried out in different vessel. This process can produce ethanol with high productivity in comparison with SHF process.

Distillation and Purification

A distillation process is necessary for separation of ethanol from mixture and purification of ethanol after fermentation process. Process is performed simply with boiling ethanol-water mixture. Because of boiling point of water (100°C) is higher than boiling point of ethanol (78 °C), ethanol vaporized before water. However, due to being an azeotrop mixture, high amount of energy is used for distillation. In order to separate azeotrop mixtures, an agent which changes the azeotrop structure must be added to the mixture. Added substance changes the volatility of mixture by effecting the molecular attractions in the mixture. Various separation agents such as benzene, pentane, cyclohexane, hexane, acetone, and diethyl ether can be used in this process. Distillation column which has two streams as top and bottom, separates most of the bioethanol from the mixture. While top stream is rich in bioethanol, bottom stream is rich in water. 37% bioethanol then concentrated in rectifying column to approach concentration of 95%. Product which is remained in the bottom is fed to stripping column in order to remove excess water. Mostly in plants, recovery of bioethanol in distillation columns is fixed to be 99.6% due to decrease bioethanol loss.

Production of Biogas (bio-methane)

Absorption

The separation principle of absorption is based on differing solubilities of various gas components in a liquid scrubbing solution. In an upgrading plant using this technique the raw biogas is forced into intensive contact with a liquid. A scrubbing column is filled with a plastic packing in order to increase the contact area between the phases. The components to be removed from the biogas (mostly carbon dioxide) are typically far more soluble in the applied liquid than methane and are removed from the gas stream. As a result, the remaining gas stream is enriched with methane and the scrubbing liquid leaving the column is rich in carbon dioxide. In order to maintain the absorption performance, this scrubbing liquid has to be replaced by fresh liquid or regenerated in a separate step (desorption or regeneration step). Currently, three different upgrading technologies embodying this physical principle are available.

Physical Absorption: Pressurised Water Scrubbing

The absorbed gas components are physically bound to the scrubbing liquid, in this case water. Carbon dioxide has a higher solubility in water than methane and will therefore be dissolved to a higher extent, particularly at lower temperatures and higher pressures. In addition to carbon dioxide, hydrogen sulphide and ammonia can also be reduced in the biomethane stream using water as a scrubbing liquid. The effluent water leaving the column is saturated with carbon dioxide and is transferred to a flash tank where the pressure is abruptly reduced and the major share of the dissolved gas is released. As this gas mainly contains carbon dioxide, but also a certain amount of methane (methane is also soluble in water, but to a smaller extent) this gas is piped to the raw biogas inlet. If the water is to be recycled back to the absorption column, it has to be regenerated and is therefore pumped to a desorption column where it meets a counter current flow of stripping air, into which the remaining dissolved carbon dioxide is released. The regenerated water is then pumped back to the absorber as fresh scrubbing liquid.

The drawback of this method is that the air components – oxygen and nitrogen – are dissolved in the water during regeneration, and thus are transported to the upgraded biomethane gas stream. Therefore, biomethane produced with this technology always contains oxygen and nitrogen. As the produced biomethane stream is also saturated with water, the final upgrading step typically is gas drying, for example by the application of glycol scrubbing.

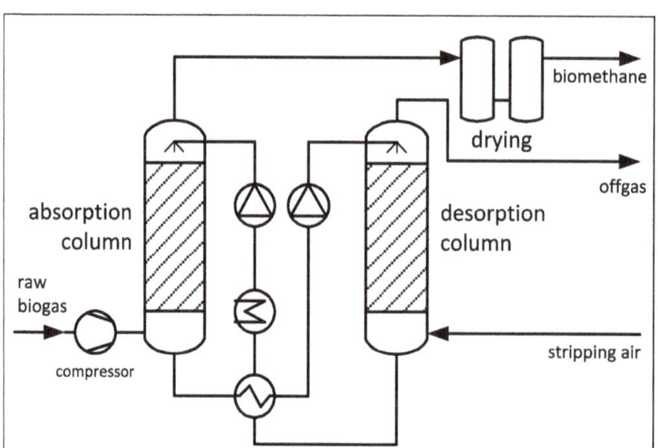

Flowsheet of a typical biogas upgrading unit applying pressurised water scrubbing; picture of the upgrading plant Könnern, Germany with a raw biogas capacity of 1,250m³/h.

Organic Physical Absorption

Using a basic similar concept to water scrubbing, this technology uses an organic solvent solution (e.g. polyethylene glycol) instead of water as a scrubbing liquid. Carbon dioxide shows higher solub-ilities in these solvents than in water. As a result, less scrubbing liquid circulation and smaller apparatus are needed for the same raw biogas capacity.

Examples of commercially available biogasupgrading technologies implementing organic physical scrubbing are Genosorbâ, Selexolâ, Sepasolvâ, Rektisolâ and Purisolâ.

Chemical Absorption: Amine Scrubbing

Chemical absorption is characterised by a physical absorption of the gaseous components in a scrubbing liquid, followed by a chemical reaction between scrubbing liquid components and ab-sorbed gas components within the liquid phase. As a result, the bonding of unwanted gas comp-onents to the scrubbing liquid is significantly stronger, and the loading capacity of the scrubbing liquid is several times higher. The chemical reaction is strongly selective and the amount of methane also absorbed in the liquid is very low, resulting in very high methane recovery and very low methane slip. Due to the high affinity of especially carbon dioxide to the solvents used (mainly aqueous solutions of Monoethanolamine MEA, Diethanolamine DEA and Methyldiethanolamine MDEA), the operating pressure of amine scrubbers can be kept significantly lower compared to pressurised water scrubbing plants of similar capacity.

Typically, amine scrubbing plants are operated at the slightly elevated pressure already arising in the raw biogas, and no further compression is needed. The high capacity and high selectivity of the amine solution, although an advantage during absorption, turns out to be a disadvantage during the regeneration of the scrubbing solution. Chemical scrubbing liquids require a significantly increased amount of energy during regeneration which has to be provided as process heat. The loaded amine solution is heated up to about 160°C where most of the carbon dioxide is released and leaves the regeneration column as a considerably pure offgas stream. As a small part of the scrubbing liquid is lost to the produced biomethane due to evaporation, it has to be replenished frequently. Hydrogen sulphide could also be absorbed from the raw biogas by chemical absorption, but even higher temperatures during regeneration would be needed. That is why it is advisable to remove this component prior to the amine scrubber.

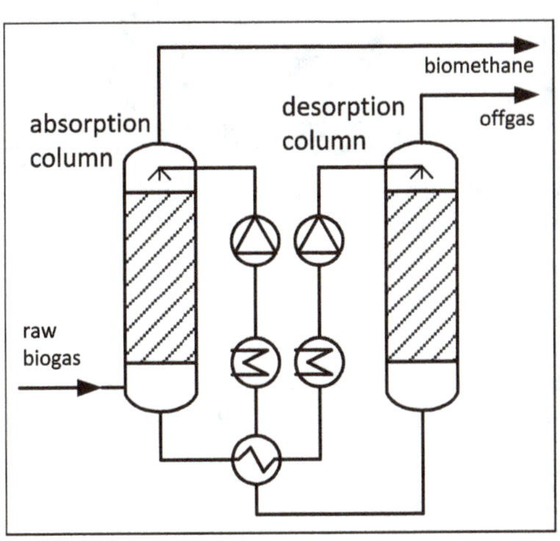

Flowsheet of a typical biogas upgrading unit applying amine scrubbing; picture of the upgrading plant Gothenburg, Sweden with a raw biogas capacity of 1,600m³/h.

Adsorption: Pressure Swing Adsorption (PSA)

Gas separation using adsorption is based on different adsorption behaviour of various gas components on a solid surface under elevated pressure. Usually, different types of activated carbon or molecular sieves (zeolites) are used as the adsorbing material. These materials selectively adsorb carbon dioxide from the raw biogas, thus enriching the methane content of the gas. After the adsorption at high pressure the loaded adsorbent material is regenerated by a stepwise decrease in pressure and flushing with raw biogas or biomethane. During this step offgas is leaving the adsorber. Afterwards, the pressure is increased again with raw biogas or biomethane and the adsorber is ready for the next sequence of loading. Industrial scale upgrading plants implement four, six or nine adsorber vessels in parallel, at different positions within this sequence in order to provide a continuous operation. During the decompression phase of the regeneration, the composition of the offgas is changing as the co-adsorbed methane is released earlier (at higher pressures) and the bulk of carbon dioxide is preferentially desorbed at lower pressures. Thus, the offgas from the first decompression stages is typically piped back to the raw biogas inlet in order to reduce the methane slip. Offgas from later stages of regeneration could be led to a second stage of adsorption, to the offgas treatment unit or could be vented to the atmosphere. As water and hydrogen sulphide contents in the gas irreversibly harm the adsorbent material, these components have to be removed before the adsorption column.

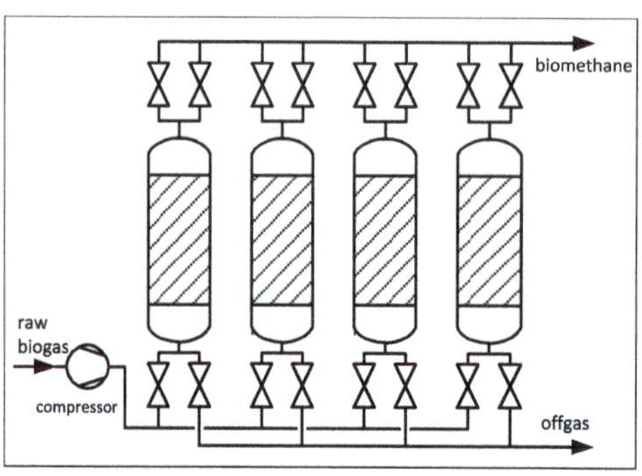

Flowsheet of a typical biogas upgrading unit applying pressure swing adsorption; picture of the upgrading plant Mühlacker, Germany with a raw biogas capacity of 1,000m³/h.

Membrane Technology: Gaspermeation

Membranes for biogas upgrading are made of materials that are permeable for carbon dioxide, water and ammonia. Hydrogen sulphide, oxygen and nitrogen permeate through the membrane to a certain extent and methane passes only to a very low extent. Typical membranes for biogas upgrading are made of polymeric materials like polysulfone, polyimide or polydimethylsiloxane. These materials show favourable selectivity for the methane/carbon dioxide separation combined with a reasonable robustness to trace components contained in typical raw biogases. To provide sufficient membrane surface area within compact plant dimensions, these membranes are applied in the form of hollow fibres, combined to assemble a number of parallel membrane modules.

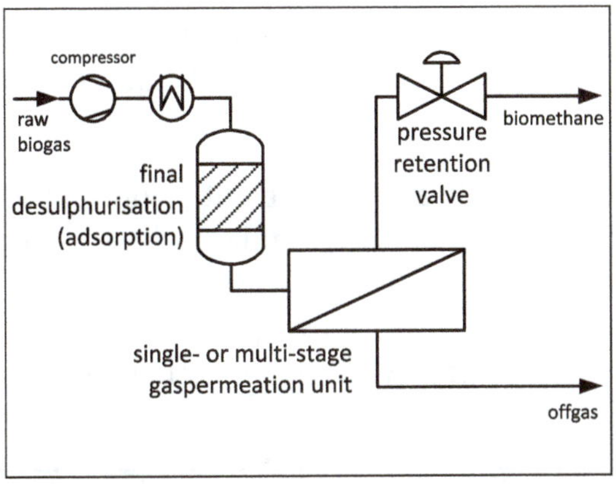

Flowsheet of a typical biogas upgrading unit applying the membrane technology gaspermeation; picture of the upgrading plant Kisslegg, Germany with a raw biogas capacity of 500m³/h.

After the compression to the applied operating pressure the raw biogas is cooled down for drying and removal of ammonia. After reheating with compressor waste heat the remaining hydrogen sulphide is removed by means of adsorption on iron or zinc oxide. Finally, the gas is piped to a single- or multi-staged gas permeation unit. The numbers and interconnection of the applied membrane stages are not determined by the desired biomethane quality, but by the requested methane recovery and specific compression energy demand. Modern upgrading plants with more complex design offer the possibility of very high methane recoveries and relatively low energy demand. Even multi-compressor arrangements have been realised and proved to be economically advantageous. The operation pressure and compressor speed are both controlled to provide the desired quality and quantity of the produced biomethane stream.

Comparison of Different Biogas Upgrading Technologies

It is hard to give a universally valid comparison of the different biogas upgrading technologies because many essential parameters strongly depend on local circumstances. Furthermore, the technical possibilities of a certain technology (for example regarding the achievable biomethane quality) often do not correspond with the most economic operation. The technical development maturity of most biogas upgrading methods nowadays is typically sufficient to meet any needs of a potential plant operator. It is mainly a question of finding a plant design providing the most economic operation for biomethane production. As a result, it is strongly recommended to perform a detailed analysis of the specific biomethane costs to be expected and to account for all possible upgrading technologies. As a guiding tool to fulfil these tasks the "Biomethane

Calculator" has been developed during this project and will be updated every year. This tool models all relevant upgrading steps and upgrading technologies and allows for a qualified estimation of specific biomethane production costs to be expected.

The following table summarises the most important parameters of the described biogas upgrading technologies, applied to a typical raw biogas composition. Values of certain parameters represent averages of realised upgrading plants or verified data from literature.

Membrane technology offers the possibility to widely adapt the plant layout to the local circum-stances by the application of different membrane configurations, multiple membrane stages and multiple compressor variations. This is why a certain range is given for most of the parameters. The first number always corresponds to the simpler plant layout ("cheaper" and with low methane recovery) while the other higher number corresponds to a high recovery plant layout.

Parameter	Water scrubbing	Organic physical scrubbing	Amine scrubbing	PSA	Membrane technology
typical methane content in biomethane [vol%]	95-99	95-99	>99	95-99	95-99
methane recovery [%]	98	96	99.96	98	80-99.5
methane slip [%]	2.0	4.0	0.04	2.0	20-0.5
typical delivery pressure [bar(g)]	4-8	4-8	0	4-7	4-7
electric energy demand [kWhel/m³ biomethane]	0.46	0.49-0.67	0.27	0.46	0.25-0.43
heating demand and temperature level	-	medium 70-80°C	high 120-160°C	-	-
desulphurisation requirements	process dependent	yes	yes	yes	yes
consumables demand	antifouling agent, drying agent	organic solvent (non-hazardous)	amine solution (hazardous, corrosive)	activated carbon (non- hazardous)	
partial load range [%]	50-100	50-100	50-100	85-115	50-105
number of reference plants	high	low	medium	high	low
typical investment costs [€/(m³/h) biomethane]					
for 100m³/h biomethane	10,100	9,500	9,500	10,400	7,300-7,600
for 250m³/h biomethane	5,500	5,000	5,000	5,400	4,700-4,900
for 500m³/h biomethane	3,500	3,500	3,500	3,700	3,500-3,700
typical operational costs [ct/m³ biomethane]					

for 100m³/h biomethane	14.0	13.8	14.4	12.8	10.8-15.8
for 250m³/h biomethane	10.3	10.2	12.0	10.1	7.7-11.6
for 500m³/h biomethane	9.1	9.0	11.2	9.2	6.5-10.1

Removal of Trace Components: Water, Ammonia, Siloxanes, Particulates

Biogas is saturated with water vapour when it leaves the digester. This water tends to condense within apparatus and pipelines and, together with sulphur oxides, may cause corrosion. By incr-easing the pressure and decreasing the temperature water will condense from the biogas and can thereby be removed. Cooling can either be realised using the ambient surroundings (air, soil) or by active cooling (i.e., refrigeration). Water can also be removed by scrubbing with glycol or by adsorption on silicates, activated charcoal or molecular sieves (zeolites).

Ammonia is usually separated when the biogas is dried by cooling, as its solubility in liquid water is high. Furthermore, most technologies for carbon dioxide removal are also selective for the removal of ammonia. A separate cleaning step is therefore usually not necessary.

Siloxanes are used in products such as deodorants and shampoos, and can therefore be found in biogas from sewage sludge treatment plants and landfill gas. These substances can create serious problems when burned in gas engines or combustion facilities. Siloxanes can either be removed by gas cooling, by adsorption on activated carbon, activated aluminium or silica gel, or by absorption in liquid mixtures of hydrocarbons.

Particulates and droplets can be present in biogas and landfill gas and can cause mechanical wear in gas engines, turbines and pipelines. Particulates that are present in the biogas are separated by fine mechanical filters (0.01μm – 1μm).

Removal of Methane from the Offgas

The offgas produced during biogas upgrading still contains a certain amount of methane depending on the methane recovery of the applied gas separation technology. As methane is a strong greenhouse gas, it is of vital importance for the overall sustainability of the biomethane production chain to minimise the methane emissions to the atmosphere. It should be noted that the emissions of methane from biogas processing plants is restricted in most countries. Additionally, higher amounts of methane in the offgas increase the specific upgrading costs and could inhibit economic plant operation. But it is never so simple, as there is a trade-off in selecting a certain methane recovery value because a higher methane recovery always increases investment and operational costs of a certain upgrading technology. As a result, the most promising plant layout in terms of economics usually accepts a certain amount of methane left in the offgas and applies a certain treatment of the gas prior to venting it to the atmosphere.

The most common technique for removing the methane content in the offgas is oxidation (i.e., combustion) and generation of heat. This heat can either be consumed at the anaerobic digestion plant itself (as this plant often has a heat demand), it can be fed to a district heating system (if locally available) or it has to be wasted by cooling. Another possibility would be to mix the offgas with raw biogas and feed it to an existing CHP gas engine. Either way, the layout of the plant has to be planned carefully, since the offgas of a modern biogas upgrading plant seldom contains enough methane to maintain a flame without addition of natural gas or raw biogas.

Alternatively, the methane in the offgas can be oxidised by a low-calorific combustor or by catalytic combustion. A number of manufacturers already provide applicable technologies on a commercial basis. These systems provide stable combustion even at methane concentrations as low as 3% in the combustion mixture with air. The treatment of offgas containing even less methane is increasingly difficult, as not enough energy is provided during the combustion of this gas, and raw biogas or biomethane have to be added in order to reach a stable oxidation. This is why it does not always make sense to choose an upgrading technology with a methane recovery as high as possible, because you always have to deal with the offgas. The integration of the upgrading plant into the biogas production facility and the overall concept of the biomethane production site are much more important. Only a very few upgrading technologies with extremely high methane recovery rates provide an offgas that is permitted to be directly vented to the atmosphere.

Production of Aviation Biofuel

Process and Technology Status

The aviation sector requires dropin bio-jet fuels that are functionally equivalent to fossil fuel and that are fully compatible with the existing infrastructure.

Jet fuel is a high-specification fuel that must meet standards as defined in ASTM D6155 for Jet A1, or by the Ministry of Defence Standard 91-91 in the UK. Bio-jet fuel must meet these same standards and in addition have an ASTM D7566 certification.

In addition to the four pathways that certified to date, several others are in the process, such as hydrotreated depolymerised cellulosic jet and green diesel (HDRD or HRD). The ASTM certification procedure is rigorous and can take years and millions of U.S. dollars to complete.

The four certified pathways to produce bio-jet fuels are described here and shown in figure below:

- HEFA: oleochemical conversion processes, such as hydroprocessing of lipid feedstocks obtained from oilseed crops, algae or tallow.

- FT: thermochemical conversion processes, such as the conversion of biomass to fluid intermediates (gas or liquid) followed by catalytic upgrading and hydroprocessing tohydrocarbon fuels.
- SIP: biochemical conversion processes, such as the biological conversion of biomass (sugars, starches or lignocellulose-derived feedstocks) to longer chain alcohols and hydrocarbons.
- ATJ: A fourth category includes "hybrid" thermochemical or biochemical technologies; the fermentation of synthesis gas; and catalytic reforming of sugars or carbohydrates.

For technologies of any type, a technology readiness- level scale is often used to determine the maturity of a technology and its closeness to market. The Commercial Aviation Alternative Fuels Initiative's fuel readiness level (FRL) scale performs this function for bio-jet pathways. The scale rates ASTMcertified fuels at FRL7 or higher.

A recent study by Mawhood evaluated pathways to bio-jet against this scale and concluded that HEFA was at FRL9, FT at FRL7 to FRL8, SIP between FRL5 and FRL7, and others in a range from four to six. Based on this scale, the recent ASTM certification of ATJ would give it a rating of FRL7. What these ratings do not capture, however, is the commercial viability of a certain pathway. As noted earlier, FT bio-jet was certified based on using coal as the feedstock, not biomass. While the feedstock does not affect fuel properties, it does have an impact on production potential, production.

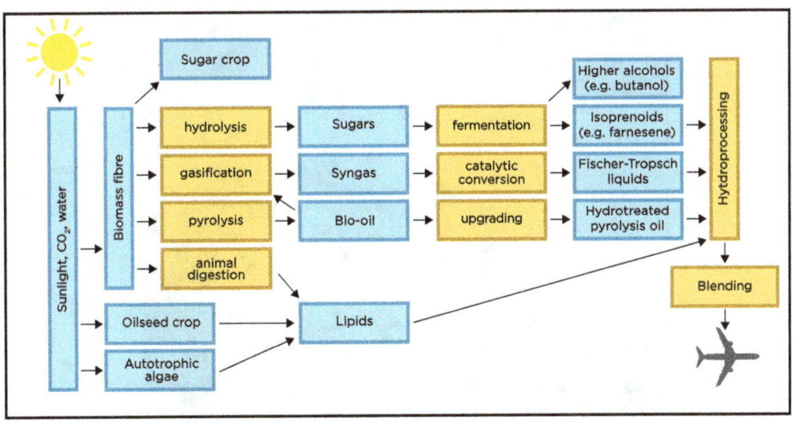

A simplified schematic diagram of different technology pathways to bio-jet fuel cost and technology readiness (RAND Corporation and Massachusetts Institute of Technology, 2009). It will also significantly impact the potential of the fuel to reduce GHG emissions. This leaves questions about the FT pathway, for which certification used coal as a feedstock. This is the closest to commercialisation among the advanced pathways based on the FRL classification, but significant challenges still must be resolved if the method is to be useful in meeting emissions-reduction targets. A recent report by France's Académie des Technologies and Académie de l'Air et de l'Espace concluded that vegetable oil-based HEFA bio-jet is likely to be the only economically viable option in the near future.

A key aspect of bio-jet production is the requirement for hydrogen (H2) to upgrade oxygen-rich carbohydrate, lignin or lipid feedstocks to hydrogenrich hydrocarbons that are functionally equivalent to petroleum-derived jet fuel. Thus, some type of hydroprocessing step will likely be required for most bio-jet fuel technology platforms, with external sources of hydrogen used to remove oxygen in the form of water from the starting material, or to saturate double bonds in a final polishing step. Processes that do not require hydrogen can be used, including chemical and biological processes, but yields will be smaller because of the consumption of a portion of the feedstock.

The amount of hydrogen needed to produce bio-jet from a feedstock is illustrated by the effective hydrogen to carbon ratio, Heff/C, in the biomass feedstocks. This hydrogen-to-carbon ratio provides a useful metric to better understand and compare the technical and economic challenges of the various drop-in biofuel processes using different types of biomass feedstocks. This staircase approach, which involves assessing how much oxygen needs to be displaced by hydrogen, establishesa ranking of feedstocks by how "easy" they are to upgrade. This shows that oils and fats need the least hydrogen, and are therefore the easiest, whereas sugars and lignocellulosic biomass need the most.

The amount required is important to determine the overall impact on GHG emissions because most hydrogen is produced by the reforming of natural gas, making fossil-fuel consumption a part of the process.

Technologies for Oloeochemical Conversion Processes

The dominant HEFA pathway uses oil and fat feedstocks such as palm oil, used cooking oil and tallow. These "fatderived" or "oleochemical-derived" drop-in biofuels are often referred to as HEFA, but are also called hydrotreated vegetable oil biofuels (HVO). This technology is mature and currently operates at a commercial scale.

HEFA is notably distinct from fatty acid methyl ester (FAME) biodiesels, which retain an oxygen ester and are therefore too oxygenated to be used as a drop-in biofuel.

Almost all of the world's commercial plants that are able to make HEFA are currently producing mostly HEFA diesel; but as of 2016, California's AltAir Fuels had become the first dedicated HEFA biojet production facility. Notably, AltAir's production process results in a mixture of hydrocarbon molecules from which several products, such as renewable diesel, bio-jet and naphtha, must be separated.

The two main technologies used commercially to make HEFA are Neste's NEXBTL and UOP and Eni's EcofiningTM processes. The EcofiningTM technology has been licensed by several companies and is used in a number of facilities.

Vegetable oils contain about 10 wt% oxygen, which must be removed to produce drop-in HEFA biofuels. Hydrotreating of vegetable oils to remove this oxygen typically consumes about 3 wt% of hydrogen. Alternative deoxygenation processes that require less or no hydrogen also produce less HEFA because of carbon losses as part of the process.

Trials have shown that HVO biofuels can also be produced by co-processing oleochemical feeds with petroleum.

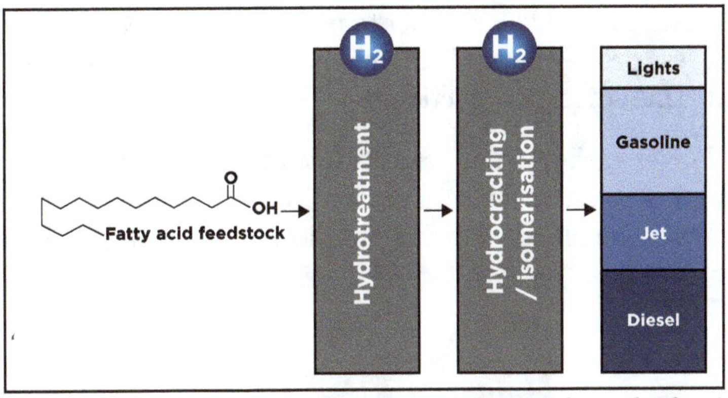

Basic Diagram of the Oleochemical Conversion Pathway feeds in modern oil refineries, although challenges remain. Chevron, Phillips66 and BP have an ASTM D7566 application underway to certify jet fuel produced through the co-processing of vegetable oils using this method at a 5% concentration.

The vast majority of the drop-in biofuels in aviation trials have used HEFA biojet primarily because the technology is mature and ASTM approved, but unless there are incentives to offset the high price, oil- and fat-based feedstocks are more likely to be converted to conventional FAME biodiesel. Another obstacle to scaling up the HEFA pathway is that increasing the fraction of jet-fuel range products produced from oleochemical feedstocks requires higher hydrogen inputs (more extensive hydrocracking) and also results in lower yields. These extra costs must be considered when developing strategies to promote increased bio-jet production. The proposed use of HEFA diesel as a bio-jet blend will minimise additional processing costs.

Vegetable-oil feedstocks require more land to produce than other feedstock types. Algae and non-food crops such as camelina, grown as a rotation crop, are examples of oleochemical feedstocks that require less land. These have been assessed by various initiatives, such as the U.S. Department of Agriculture "farm-to-fly" program; the U.S.

Department of Energy's National Alliance for Advanced Biofuels and Bioproducts consortium, which is now complete; the European 7th Programme for Research and Innovation; Horizon 2020; and the "Flightpath" programs. Cooking oil and tallow are much more sustainable, but availability is limited. In the short- to mid-term neither have the potential to contribute significantly to any increase in the production of bio-jet. The U.S. Environmental Protection Agency has estimated that about 3 bln gallons (11.4 bln L) of used cooking oil is produced each year in the U.S., resulting in a theoretical yield of about 9 bln L of HEFA bio-jet, based on 1.2:1 feedstock to fuel ratio. However, much of this feedstock is currently used to make biodiesel.

One strategy for improving costs of bio-jet based on the oleochemical conversion route is the use of HDRD diesel as a jet substitute. This also involves the upgrading of vegetable oils. Boeing and Neste have applied for ASTM certification under ASTM D7566 to blend small amounts of vegetable oil with ordinary jet fuel20. Boeing completed a test flight using a 15% blend of HDRD in December 2014. Although no information is available on costs, this blend should be cheaper to produce than HEFA bio-jet, as the blend is likely to require less processing and result in potentially higher yields. The process is similar to current HEFA diesel production.

It differs in that longer carbon-chain lengths from vegetable oils are not cracked into shorter jet-range molecules. Instead increased isomerisation is used to improve the cold flow properties of the blend, allowing low bio-jet blends that still meet the overall specifications for jet fuel.

Thermochemical Routes to Turn Biomass Into bio-jet fuel

The major challenges for thermochemical routes to bio-jet differ mainly in conversion-process efficiency and technology risks, although feedstock choices can also result in qualities in the end product. Thermochemical routes used to turn biomass into bio-jet involve the production of three main products, in different ratios: bio-oil, synthesis gas and char. The two main thermochemical routes to bio-jet are gasification and pyrolysis, and hydrothermal liquefaction (HTL). The FT process uses gasification combined with synthesis to produce bio-jet. Several commercial facilities based on gasification-FT are planned, and this pathway is discussed in more detail below. The pyrolysis route to biojet is known as HDCJ (hydrotreated depolymerised cellulosic jet). An ASTM application for HDCJ was initiated by KiOR, but the company is now in bankruptcy, creating a setback for the certification of this pathway.

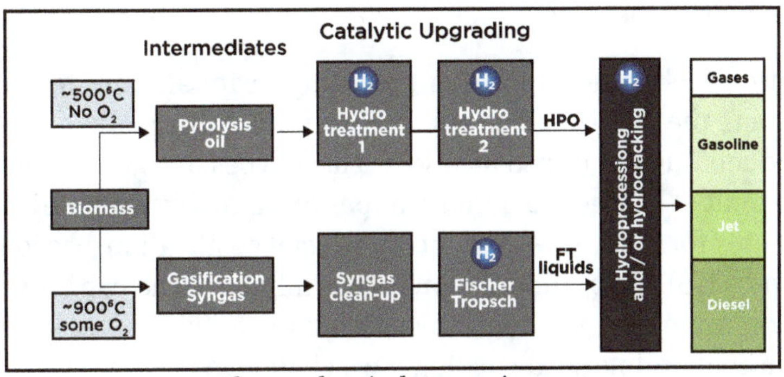

Thermochemical conversion routes.

Gasification

Gasification involves the heating of small feedstock particles at high temperatures in a controlled-oxygen environment to produce synthesis gas, which is comprised of mostly H_2 and carbon monoxide and typically called syngas. Syngas converts to numerous gaseous and liquid chemicals or fuels via the FT process using catalysts. This process produces a mixture of hydrocarbon molecules from which various fuels and chemicals can be extracted.

Gasification and FT have been used since the 1980s by South Africa's Sasol company to convert coal into fuels, at a current capacity of 160 000 barrels per day (b/d). The FT process is also used in the world's largest natural gas-to-liquids plant, Shell's Pearl facility in Qatar. It was completed in 2011 and produces 140 000 b/d of fuels (IEA Bioenergy, 2014). Although bio-jet could potentially be produced via this process, using biomasses such as black liquor or bio-oil as a feedstock, various challenges have prevented this on a commercial basis.

Bio-jet produced via FT was certified by ASTM in 2009, at up to 50% of a blend with conventional jet fuel, but it was produced using coal as the feedstock and not biomass. The feedstock that is used to make FT syngas can result in different products during and after gasification. This significantly influences the composition of the syngas. For efficient fuel production, a desirable syngas should be a mixture of H2 and CO only, with all contaminants removed.

Gasification of biomass typically results in considerable tar formation that needs to be cleaned up, and the high oxygen content of biomass impacts the ratio of H_2 to CO in the synthesis gas. As a result, biomass-derived syngas is less energy dense than syngas derived from natural gas, has a lower ratio of hydrogen to carbon, and contains more impurities. Typically, biomass and MSW-derived syngas needs to be enriched in hydrogen and cleaned of impurities such as tars, nitrogen and other atoms comprised of anything other than carbon or hydrogen. These impurities can deactivate the synthesis catalysts. Although cleaning syngas is technically possible, it has proven to be costly. Plasma gasification is another way to produce a very clean syngas, but has also proven to be significantly more expensive.

To date, gasification technologies have entailed high capital costs to both gasify the biomass and convert the resulting syngas to FT liquids or partially oxygenated liquid hydrocarbon products such as mixed alcohols. These types of coalor natural gas-fed facilities have been built at a large scale in the hopes of capturing economies of scale. The capital cost estimates for a first-of-itskind commercial gasification-based facility range from USD 600 m to USD 900 m, and would typically have the capacity to produce 2 000 t per day of dry biomass. Although this is a significantly smaller size than current facilities based on coal and natural gas, logistics challenges will be likely because biomass is a less energy-dense feedstock.

Some of these anticipated supply-chain challenges can be mitigated through the use of alternative feedstocks such as pyrolysis bio-oils, which are more energy dense than wood. Because a range of hydrocarbon molecules are produced by the FT process, large-scale facilities are also in a better position to market the multiple commercial products created, thus improving the overall economics.

Current FT technology results in a maximum of about 40% of the final product comprised of bio-jet fuel and middle distillates, requiring the marketing of the other 60% of the output.

Commercial biomass-gasification facilities under construction include those of Fulcrum Bioenergy and Red Rock Biofuels, in U.S. Kaidi has proposed building an FT facility in Finland. These pioneer plants should provide invaluable insights and lessons for future investment. Many people believe that costs for the FT route could fall considerably as the technology matures. Fulcrum Bioenergy claims to be able to produce FT transportation fuel at less than USD 1 per gallon (USD 0.26/L) using MSW as a feedstock.

A proposal from Solena and British Airways to use MSW in a gasification process was recently cancelled. Solena hoped to use plasma gasification technology that was likely to cost significantly more than what Fulcrum will use. Red Rock Biofuels plans to use woody biomass as well as a different FT technology (from Velocys).

Pyrolysis and Hydrothermal Liquefaction (HTL)

Fast pyrolysis exposes small biomass particles of about 3 mm in length to heat at 500°C for a few seconds to produce a bio-oil with up to 75 wt% yield. Although companies such as Ensyn in Canada have been producing fast-pyrolysis biooils for many years, these have mainly been used in niche applications such as food flavouring.

Energy applications have been restricted to heavy fuel oil used in stationary heating, and in powergenerating facilities. Although Ensyn22 recently obtained regulatory approval for RFDiesel and RFGasoline, which are fuel products generated via co-processing in petrochemical refineries, no jet fuel has been produced this way.

In the Netherlands, BTG has commercialised the flash pyrolysis technology in its EMPYRO23 project. As of late 2016, however, biooil was used to replace natural gas in a heating application in a milk factory. BTG has also been testing possible co-processing of bio-oil in a petroleum refinery.

The commercial production of bio-jet via the pyrolysis route is likely to be challenging because biocrudes derived from fast pyrolysis contain up to 40% oxygen, similar to the biomass itself. This necessitates extensive upgrading to produce bio-jet, which is typically achieved through hydroprocessing. These processing costs, as well as the need for external hydrogen, represent a large proportion of equipment and production costs.

A further challenge to the hydroprocessing of pyrolysis oils is the cost and stability of the catalysts that are required.

A potential advantage of the pyrolysis approach to bio-jet production is that it can be done in existing oil refineries, which reduces the need for capital to build a dedicated facility.

Similarly, significant savings might be achieved by directly sourcing hydrogen from an oil refinery and, in the longer term, through using existing processing units.

Co-processing in existing petroleum refineries is considered a key strategy for upgrading pyrolysis-derived bio-oils, but comes with some technical challenges. These include selecting the point of insertion, the extent to which upgrading is required prior to insertion and the disparate types of catalysts needed for bio-oils compared with those used in oil refining. Refinery-insertion strategies should be synergistically beneficial but are likely more technically challenging than is generally acknowledged.

Catalytic pyrolysis or processes such as HTL can produce a bio-oil intermediate with significantly lower oxygen content, at less than 10%. That would be easier to upgrade to produce fuels, including bio-jet.

Although some studies have indicated that this method could potentially produce the lowest-cost bio-jet, the high-pressure requirements of HTL during the production of biocrude will impact their potential for scale-up. While production of bio-oil via pyrolysis is at a commercial scale, HTL is currently just at the demonstration stage, as pioneered by Licella's Australian plant. Although there is a scarcity of reliable technical and economic analyses, a minimum fuel-selling price (MFSP) of USD 3.39 per gallon could be achieved when making diesel and gasoline via fast pyrolysis followed by upgrading.

Biochemical Routes to Turn Biomass Into Bio-jet Fuel

Capital expenditure for biochemical routes to bio-jet are projected to be lower than those for thermochemical routes, but that benefit may be offset if the MFSP is higher, which is also expected (table). In contrast to the more familiar sugar-to-ethanol fermentation route to bioethanol, advanced biological routes convert sugars to less oxygenated and more energydense molecules such as longer-chain alcohols like butanol and butanediol. Advanced biological routes can also convert sugars to larger hydrocarbon molecules such as isoprenoids and fatty acids. Amyris, using genetically engineered yeasts, converts sugars directly to renewable hydrocarbons such as farnesene. Farnesene can then be upgraded to farnesane through hydroprocessing, producing SIP. It is also known as the Direct Sugars to Hydrocarbons (DSHC) pathway. It received ASTM certification in 2014, provided it is used in a 10% blend with fossil-derived jet fuel.

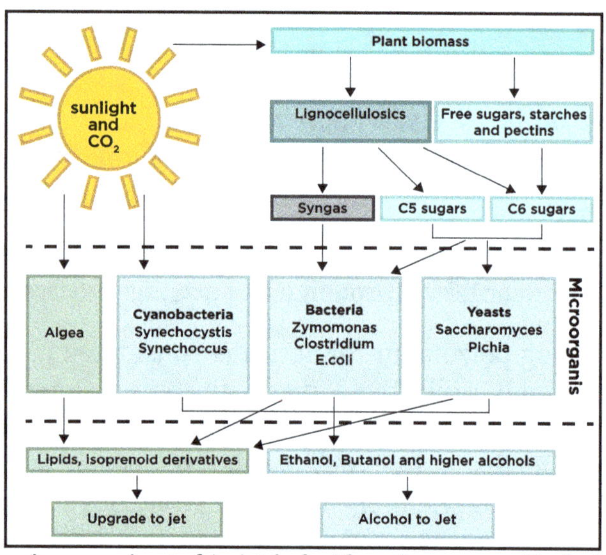

From microorganism to bio-jet fuel with various renewable feedstocks.

Although butanol, n-butanol and isobutanol are oxygenated and thus cannot be considered to be fully "dropin" biofuels, they are less oxygenated and less hydrophilic than ethanol and can be used to produce bio-jet fuel through the ATJ pathway. Gevo recently obtained ASTM certification for its bio-jet fuel produced from isobutanol, which Alaska Air has used in a commercial flight.

The potential to use existing ethanol facilities lies in the ability to replace the existing yeasts that produce ethanol with alternative microorganisms that could instead make bio-jet. Because some of the biological intermediates such as farnesene are quite hydrophobic they should in theory be more readily recoverable from the aqueous fermentation broths. Nevertheless, the recovery of these molecules from the fermentation broth has been more challenging than predicted because of intracellular expression of hydrophobic metabolites. For butanol, fermentation titers are typically well below the concentration that induces phase separation, which ranges from 70 g/L of butanol to 80 g/L.

Biochemical-based bio-jet pathways can produce highly reduced biohydrocarbon molecules such as sesquiterpenes and fatty acids based on the used microorganism, which accordingly decrease the degree of final hydroprocessing required to meet jet-fuel specifications. Even still, advanced biological pathways requiremore energy and carbon-intensive metabolic processes than required in ethanol production in order to decrease feedstock sugars.

Given the potential commercialisation scale, conventional ethanalogenic yeast fermentations could achieve a higher order of magnitude than the product yield for fermentation-derived biojet. Furthermore, biochemical-based bio-jet platforms are relatively pure, and functionalised long carbon-chain molecules can be produced. This is an advantage over thermochemical-based processes.

The production of bio-jet would not be the most profitable use of biochemically processed biomass and sugars. Products such as carboxylic acids, alcohols and polyols can generate higher profits because there are fewer processing steps, lower NADPH requirements, and less hydrogen consumption. Less-oxygenated microbial metabolites with potential as drop-in biofuel intermediates are already being sold in the value-added chemicals and cosmetics markets, such as Amyris's farnesene and Gevo's or Butamax's butanol. The market for biochemical drop-in products is highly competitive and growing. The volume of biochemical drop-in products in the market is expected to increase by between 10 Mt and 50 Mt annually to 2020, which is equal to the current markets size of biofuels. Incentives will be required for commercial entities to concentrate on drop-in fuels until the market is saturated.

Hybrid Conversion Processes

Hybrid conversion processes can combine a mixture of the above methods. For example, the ATJ process can combine biochemical production of alcohol, through fermentation of sugars or conversion of syngas from gasification and catalytic conversion of the alcohol to bio-jet. Another example of a hybrid process is Virent's aqueous phase reforming. This process uses feedstock sugars potentially derived through biochemical conversion of lignocellulosic feedstocks, while synthesis takes placevia catalysis.

Methods used in Production

Supercritical Process

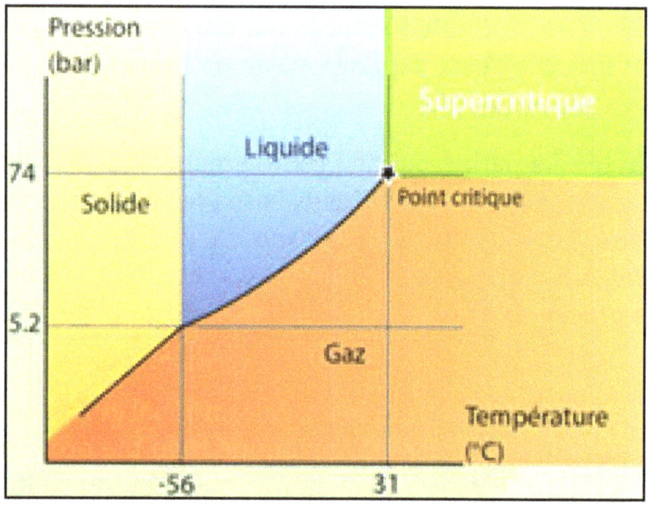

Any substance is characterized by a critical point which is obtained at specific conditions of pressure and temperature. When a compound is subjected to a pressure and a temperature higher than its critical point, the fluid is said to be "supercritical".

In the supercritical region, the fluid exhibits particular proporties and has an intermediate behavior between that of a liquid and a gas. In particular, supercrical fluids (SCFs) possess liquid-like densities, gas-like viscosities and diffusities intermediate to that of a liquid and a gas.

As shown on the phase diagram, the gas-liquid equilibrium curve is interrupted at the critical point, providing a continuum of physico chemical properties.

The fluid is said "supercritical" when it is heated obove its crititcal temperature and compressed obove its cirical pressure.This particular behavior of substances was first observed in 1822 by French engineer and physicist, Charles Cagniard de La Tour in his famous cannon barrel experiment. It was then defined as supercritical fluid by Irish chemist, Thomas Andrews. The most widely used supercritcal fluids are CO_2 and water.

Supercritical fluids based processes include extraction, impregnation, particle formation, formulation, sterilization, cleaning and chemical reactions among others. In all cases, the supercritical fluid is used as an alternative to traditional organic liquid solvents.The most widely used supercritical fluids are CO_2 (Tc = 31°C, Pc = 74 bar) and water (Tc = 374°C, Pc = 221 bar) but some processes (extraction, reactions) involove the use of supercritical methanol, ethanol, propane, ethane mainly. Supercritical CO_2 processes are the most widely spread as these are exempt of the operations of elimination of solvent residues, operations generally needed when the solvent used is an organic compound.

Some examples of proceseses carried out under supercritical and subcritical conditions are listed below.

Extraction

The principle of the extraction of solids using supercritical CO_2 relies on the strong variation of the solvation power of CO_2 which occurs by simple variation the operating conditions (temperature and pressure). This allows to selectively extract molecules according to their chemical nature. Very weakly polar, CO_2 turns out to be an excellent solvent of nonpolar or small polar molecules in supercritical conditions. Once the desired compound is dissolved in supercritical CO_2, the pure extract can easily be obtained by simple depressurization. This results in the separation of CO_2, which becomes gas again, from the extract, which is recovered in solid or liquid form. In theory, any porous solid material (plants, plastics, wood, can be treated by supercritical fluid extraction to recover valuable compounds (oils, fragrances, pigments) or undesired substances (pollutants, residual solvents).

Subcritical water extraction can also be carried out to extract hydrophobic compounds (polyphenols, terpenes, tannins) from various plant materials (fruits, wood, algae).

Fractionation

Based on the principle of supercritical CO_2 extraction of solids, the extraction (fractionation) of liquids can also be performed in order achieve high levels of purification. Examples of liquids concerned by this process include plant extracts, vegetable oils, fish oil, polymers.

Impregnation

The principle of supercritical fluid impregnation consists of a scan of a porous solid material (polymers, wood, textiles) by a supercritical (mainly CO_2) in which the active substance to impregnate is previously dissolved. This step is followed by a relaxation that causes the passage of the CO_2 in the gaseous state, leaving the "target" material impregnated with the active substance. Therefore, the impregnation of a solid matrix with an active compound is easy achievable using supercritical fluids based technologies. Example of the use of supercritcal impregnation include the dyeing of textile, the tanning of leather, wood impregnation.

Powder Formation

This refers to micro and nanoparticle generation, crystalization, precipitation, micronization of inorganic, organic, pharmaceutical, and polymeric materials.There are two main ways of precipitating micro and nanoparticles either using supercritical fluid as solvent, the RESS technique (Rapid Expansion of Supercritical Solutions); or using it as anti-solvent, the SAS technique (Supercritical Anti-Solvent). Encapsulation of variuous substances and polymers can also be performed using the PGSS (Particles from Gas Saturated Solutions) process.

Deposition

Chemical Fluid Deposition (CFD) involves the chemical reduction of organometallic compounds in supercritical fluids to yield high purity deposits. Typically, the reaction is initiated upon the addition of H_2 or other reducing agent. The advantages of CFD over conventional deposition techniques are a consequence of the unique properties of supercritical fluids, thus promoting infiltration into complex geometries and minimizing mass transfer limitations common to liquid phase reductions. High purity metal films including Pt, Pd, Au, Rh, Ni, Cu, Al have been deposited by CFD from supercritical fluids CO_2 using appropriate precursors.

Chemical Reactions

Conducting chemical reactions at supercritical conditions affords opportunities to manipulate the reaction environment (solvent properties) by increasing pressure to enhance the solubilities of reactants and products, to eliminate interphase transport limitations thus increasing reaction rates, and to integrate reaction and separation unit operations. Supercritical conditions may be advantageous for reactions involued in fuels processing, biomass conversion, biocatalysis, homogeneous and heterogeneous catalysis, environmental control, polymerization, materials synthesis and chemical synthesis. Examples of chemical reactions carried out at an industrial scale include hydrogenation, oxydation, esterification and etherfication reactions among others.

Sterilization

Supercritical fluids have demonstrated the ability to inactivate bateria, fungi, yeasts and viruses thus providing an efficient mean for the sterilization of food and medical devices. The mechanism of micro-organism inactivation are not fully understood yet but some evidence show that this may be caused by cell wall alteration due to a strong interaction of the fluid with the lipids and the inactivation of key-enzymes resulting from pH decrease inside the cell.

Drying

Supercritical drying processes rely on the extraction of water and other solvents using supercritical CO_2. The absence of surface tension allows the supercritical fluid to be removed without distortion. Such processes are used to make aerogels but also for the lyophilization (freeze-drying) of biological and food matrices and the dry-cleaning of clothes (as a replacement for chlorinated solvenst) in an environmentally-friendly way.

Cleaning

In the cleaning process, the use of supercritical CO_2 with or without the addition of specific surfactants avoids the use of solvents such as trichloroethylene. For environmental reasons, this toxic solvent is subject to many health and regulatory limitations, leaving the place to "green" solvents. Although only hydrocarbon solvents are nowadays able to meet the market's demand for alternative solutions to chlorinated solvents, the use of CO_2 under pressure or supercritical CO_2 appears as a particularly interesting industrial alternative to chlorinated solvents. This process is industrialized for the dry-cleaning of textiles, the cleaning of mechanical spare parts.

Chromatography

Supercritical Fluid Chromatography (SFC) is used for the analysis and purification of low to moderate molecular weight, thermally labile molecules and the the separation of chiral compounds. Principles are similar to those of high performance liquid chromatography (HPLC), however SFC typically utilizes carbon dioxide as the mobile phase. Therefore the technique is more versatile, exhibits better resolution and faster analysis times than conventional liquid chromatographic methods.

Refrigeration

Supercritical CO_2 can be used as a refrigerant in air-conditioning and cooling system providing eco-friendly solutions in industrial and domestic heat-pumps.

Ultra- and high-shear in-line and Batch Reactors

Ultra- and High Shear in-line or batch reactors allow production of biodiesel continuously, semicontinuously, and in batch-mode. This method drastically reduces production time and increases production volume. The reaction takes place in the high-energetic shear zone of the ultra- and high Shear mixer by reducing the droplet size of the immiscible liquids such as oil or fats and methanol. Therefore, the smaller the droplet size, the larger the surface area the faster the catalyst can react.

Ultrasonic Reactor Method

In the ultrasonic reactor method, the ultrasonic waves cause the reaction mixture to produce and collapse bubbles constantly. This cavitation provides simultaneously the mixing and heating required to carry out the transesterification process. The ultrasonic reactor method for biodiesel production drastically reduces the reaction time, reaction temperatures, and energy input. Industrial scale ultrasonic devices allow for the industrial scale processing of several thousand barrels per day.

Ultrasonic cavitation provides the necessary activation energy for the industrial biodiesel transesterification.

Transesterification

Manufacturing biodiesel from vegetable oils (e.g. soy, canola, jatropha, sunflower seed or algae) or animal fats, involves the base-catalyzed transesterification of fatty acids with methanol or ethanol to give the corresponding methyl esters or ethyl esters. Glycerin is an inevitable byproduct of this reaction.

Vegetable oils as animal fats are triglycerides composed of three chains of fatty acids bound by a glycerin molecule. Triglycerides are esters. Esters are acids, like fatty acids, combined with an alcohol. Glycerine (= glycerol) is a heavy alcohol. In the conversion process triglyceride esters are turned into alkyl esters (= biodiesel) using a catalyst (lye) and an alcohol reagent, e.g. methanol, which yields methyl esters biodiesel. The methanol replaces the glycerin.

The glycerine the heavier phase will sink to the bottom. Biodiesel the lighter phase floats on top and can be separated, e.g. by decanters or centrifuges. This conversion process is called transesterification.

The conventional esterification reaction in batch processing tends to be slow, and phase separation of the glycerin is time-consuming, often taking 5 hours or more.

Ultrasonics for Biodiesel Processing

Today, biodiesel is primarily produced in batch reactors. Ultrasonic biodiesel processing allows for the continuous inline processing. Ultrasonication can achieve a biodiesel yield in excess of 99%. Ultrasonic reactors reduce the processing time from the conventional 1 to 4 hour batch processing to less than 30 seconds. More important, ultrasonication reduces the separation time from 5 to 10 hours (using conventional agitation) to less than 60 minutes. The ultrasonication does also help to decrease to amount of catalyst required by up to 50% due to the increased chemical activity in the presence of cavitation. When using ultrasonication the amount of excess methanol required is reduced, too. Another benefit is the resulting increase in the purity of the glycerin.

Ultrasonic processing of biodiesel involves the following steps:

- The vegetable oil or animal fat is being mixed with the methanol (which makes methyl esters) or ethanol (for ethyl esters) and sodium or potassium methoxide or hydroxide.

- The mix is heated, e.g. to temperatures between 45 and 65degC.
- The heated mix is being sonicated inline for 5 to 15 seconds.
- Glycerin drops out or is separated using centrifuges.
- The converted biodiesel is washed with water.

Most commonly, the sonication is performed at an elevated pressure (1 to 3bar, gauge pressure) using a feed pump and an adjustable back-pressure valve next to the flow cell.

References

- Biodiesel-transesterification-01: hielscher.com, Retrieved 08, March 2020
- Supercritical-fluids.146.0: supercriticalfluid.org, Retrieved 26, April 2020
- Irena-biofuels-for-aviation-2017: irena.org, Retrieved 19, February 2020
- Biofuels-status-and-perspective/an-overview-of-bioethanol-production-from-algae: intechopen.com, Retrieved 23, July 2020
- Chemistry-of, biodiesel: goshen.edu, Retrieved 20, July 2020

5

Biofuel Standards and Policies

There are various standards in place which govern the production, storage, usage and transportation of different biofuels. This chapter closely examines the various biofuel standards and policies in place around the world such as biofuel policy in Europe and biofuel policy in Brazil to provide an extensive understanding of the subject.

The production and use of biofuels – ethanol and biodiesel – has experienced a remarkable growth over the last decade. According to the US Energy Information Agency (EIA) data, total world biofuel production increased nearly six-fold over the 2000- 2010 period, from 315 thousand barrels per day to 1856 thousand barrels per day. Three countries/regions have been leading this development: the United States, Brazil and the European Union (EU). Ethanol has been the leading biofuel in the Unites States (from corn) and in Brazil (from sugarcane), whereas biodiesel is the preferred biofuel in Europe (rapeseed oil is the most important feedstock). Whereas both types of biofuels have experienced a similar upward trend in recent years, ethanol remains the dominant type of biofuel. In 2010 ethanol accounted for three-fourth of world biofuel output (when expressed in energy equivalent units). Ethanol production increases have been particularly impressive in the United States, where the annual rate of growth over the period 2000-2010 was more than double that in Brazil. Indeed, the United States surpassed Brazil as the world's largest ethanol producer in 2006, and by 2010 it produced 57.1% of world's ethanol output.

Whereas the development of this industry draws on roots established long ago, especially in Brazil, its recent boom has been heavily influenced by critical policies that are promoting the production and consumption of biorenewables in general, and biofuels in particular. The rapid developments that have affected biofuel policies and biofuel industries have generated considerable debate, and a number of unresolved issues remain.

A number of countries have experienced recent rapid increases in both supply and demand of biofuels, but the US ethanol sector stands out for the rate of growth experienced in the last decade, which has made the United States the leading producer of biofuels worldwide. In the United States ethanol has been produced from corn for more than three decades. The production trend depicted in Figure below shows a slow albeit steady growth up to the beginning of the new millennium, and a markedly increased growth rate over the last decade. In 2011 ethanol output reached 13.9 billion gallons, an 80-fold increase relative to the 1980 level. Throughout this period, a few federal policies have played a key role in the development of this industry. An initial stimulus came from the $0.40/gallon subsidy (technically, an excise tax exemption) established by the Energy Tax Act of 1978. In various forms, the federal subsidy was active until it was allowed to expire at the end of 2011. It increased early on to reach $0.60/gallon with the Tax Reform Act of 1984, but has been gradually adjusted downward since 1990. The subsidy (by then a blender tax credit) was last decreased to $0.45/gallon as of January 2009, and it was finally phased out at the end of December 2011. The desire to keep the subsidy for domestic production only motivated the introduction of a $0.54/gallon duty on ethanol imports (to supplement the out-of-quota bound ad valorem import tariff of 2.5%). This specific import duty also expired at the end of 2011. Many other federal and state programs exist that provide biofuel incentives.

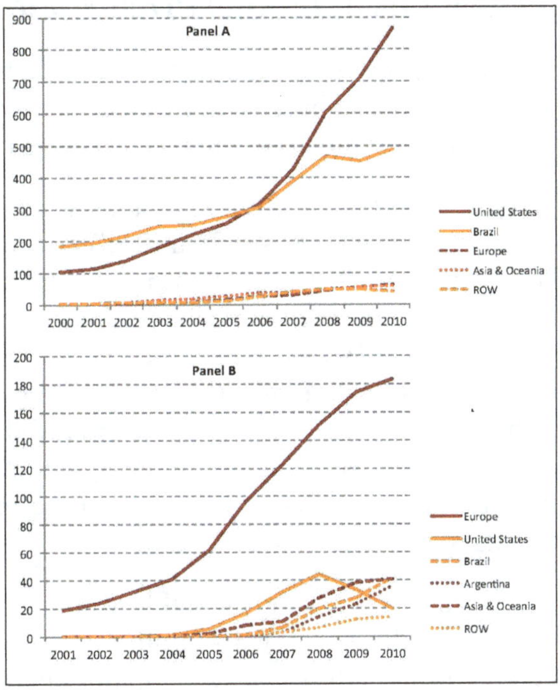

Fig: World production of biofuels (thousand barrels per day). Panel A: ethanol; Panel B: biodiesel.

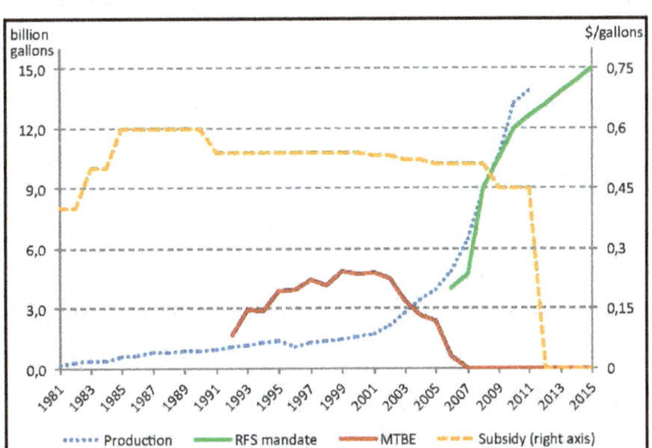

Fig: Evolution of the US Ethanol Industry.

Whereas the federal subsidy undoubtedly supported the earlier growth of the US ethanol industry, environmentally-led regulations also played an important role. In particular, the 1990 amendments to the Clean Air Act introduced a 2% oxygen requirement for gasoline. As concerns eventually arose as to the groundwater contamination potential of methyl tertiary butyl ether (MTBE), an early favorite oxygenate gasoline additive and octane enhancer, and following bans by some states (led by California), the gasoline refining industry began phasing out MTBE in the early 2000s. Ethanol emerged as the most viable oxygenate substitute for MTBE, which fostered a valuable market niche for ethanol as a key gasoline additive.

Potential Economic Benefits of Biofuel Production

Replacing fossil fuels with biofuels has the potential to generate a number of benefits. In contrast to fossil fuels, which are exhaustible resources, biofuels are produced from renewable feedstocks. Thus, their production and use could, in theory, be sustained indefinitely.

While the production of biofuels results in GHG emissions at several stages of the process, EPA's analysis of the Renewable Fuel Standard (RFS) projected that several types of biofuels could yield lower lifecycle GHG emissions than gasoline over a 30 year time horizon. Academic studies using other economic models have also found that biofuels can lead to reductions in lifecycle GHG emissions relative to conventional fuels. Second and third generation biofuels have significant potential to reduce GHG emissions relative to conventional fuels because feedstocks can be produced using marginal land. Moreover, in the case of waste biomass, no additional agricultural production is required, and indirect market-mediated GHG emissions can be minimal if the wastes have no other productive uses.

Biofuels can be produced domestically, which could lead to lower fossil fuel imports. If biofuel production and use reduces our consumption of imported fossil fuels, we may become less vulnerable to the adverse impacts of supply disruptions. Reducing our demand for petroleum could also reduce its price, generating economic benefits for American consumers, but also potentially increasing petroleum consumption abroad.

Biofuels may reduce some pollutant emissions. Ethanol, in particular, can ensure complete combustion, reducing carbon monoxide emissions.

It is important to note that biofuel production and consumption, in and of itself, will not reduce GHG or conventional pollutant emissions, lessen petroleum imports, or alleviate pressure on exhaustible resources. Biofuel production and use must coincide with reductions in the production and use of fossil fuels for these benefits to accrue. These benefits would be mitigated if biofuel emissions and resource demands augment, rather than displace, those of fossil fuels.

Potential Economic Disbenefits and Impacts of Biofuel Production

Biofuel feedstocks include many crops that would otherwise be used for human consumption directly, or indirectly as animal feed. Diverting these crops to biofuels may lead to more land area devoted to agriculture, increased use of polluting inputs, and higher food prices. Cellulosic feedstocks can also compete for resources (land, water, fertilizer, etc.) that could otherwise be devoted to food production. As a result, some research suggests that biofuel production may give rise to several undesirable developments.

Changes in land use patterns may increase GHG emissions by releasing terrestrial carbon stocks to the atmosphere. Biofuel feedstocks grown on land cleared from tropical forests, such as soybeans in the Amazon and oil palm in Southeast Asia, generate particularly high GHG emissions. Even use of cellulosic feedstocks can spur higher crop prices that encourage the expansion of agriculture into undeveloped land, leading to GHG emissions and biodiversity losses.

Biofuel production and processing practices can also release GHGs. Fertilizer application releases nitrous oxide, a potent greenhouse gas. Most biorefineries operate using fossil fuels. Some research suggests that GHG emissions resulting from biofuel production and use, including those from indirect land use change, may be higher than those generated by fossil fuels, depending on the time horizon of the analysis.

Regarding non-GHG environmental impacts, research suggests that production of biofuel feedstocks, particularly food crops like corn and soy, could increase water pollution from nutrients, pesticides, and sediment. Increases in irrigation and ethanol refining could deplete aquifers. Air quality could also decline in some regions if the impact of biofuels on tailpipe emissions plus the additional emissions generated at biorefineries increases net conventional air pollution.

Renewable Fuel Standard

Established by Congress as an amendment to the Clean Air Act, the Renewable Fuel Standard (RFS) mandates that U.S. transportation fuels contain a minimum volume of biofuel. The mandated minimum volume increases annually and must be met using both conventional biofuel (e.g., corn starch ethanol) and advanced biofuel (e.g., cellulosic ethanol). For a renewable fuel to be applied toward the mandate, it must be used for certain purposes (i.e., transportation fuel, jet fuel, or heating oil) and meet certain environmental and biomass feedstock criteria.

The statute outlines volume requirements—listed in tables—for four biofuel categories: total renewable fuel, total advanced biofuel, cellulosic biofuel, and biomass-based diesel. The total renewable fuel statutory volume required for any given year equates to the sum of conventional biofuel (which is unspecified in statute) and advanced biofuel (which is specified in statute). Both cellulosic biofuel and biomass-based diesel are subcategories of advanced biofuel (both of which are specified in statute). There is also a third advanced biofuel category—other advanced biofuels (which is unspecified in statute).

The statutory volume requirements for both total renewable fuel and total advanced biofuel for the RFS have not been met since 2013. The Environmental Protection Agency (EPA) Administrator has the authority to waive the RFS requirements, in whole or in part, if certain conditions outlined in statute prevail. EPA has used this waiver authority multiple times. For instance, the 2020 targets set by EPA for total renewable fuel and for total advanced biofuel were approximately 67% and 34% of the statutory targets, respectively. A variety of factors, such as infrastructure, technology, and limited federal assistance, have led to challenges in meeting the total volume requirement established by Congress. These challenges have included a lack of cellulosic biofuel production and delays by the EPA in approving fuel pathways.

There are, however, two fuel categories that consistently have met their statutory targets: conventional biofuel and biomass-based diesel. Also, since 2014, two advanced biofuel pathways—renewable compressed natural gas and renewable liquefied natural gas—have constituted the majority of the reduced cellulosic biofuel volume targets established by EPA.

A multitude of factors affect the conditions under which the RFS operates—some external to RFS policy and some internal. The impact of these factors can be challenging to quantify. For example, it is not clear how changes in gasoline consumption in response to fluctuating crude oil and gasoline prices and a global event like the COVID-19 pandemic impact the biofuel or conventional fuel industries. It is not clear how agricultural trade issues impact biofuel policies or the commodities used to produce biofuel. It is also uncertain how the program will fare if EPA implements the "reset" provision of the statute, which allows the agency to modify the volumes required for future years if certain conditions are met.

Challenges to implementing the RFS have led to scrutiny of the program in Congress and to litigation about EPA's regulations. Largely due to concerns about the implementation and feasibility of the RFS, some Members of Congress have expressed their perspectives on EPA's rulemakings as well as EPA's implementation of the program. They also have questioned whether to amend or repeal the RFS or whether to maintain the status quo.

The Statute

The RFS was established by the Energy Policy Act of 2005. It was expanded in 2007 by the Energy Independence and Security Act. The RFS mandate requires that transportation fuels sold or introduced into commerce in the United States contain an increasing volume of a predetermined suite of renewable fuels. The statute required 4.0 billion gallons of renewable fuel in 2006, ascending to 36.0 billion gallons required in 2022, with EPA determining the volume amounts after 2022 in future rulemakings.

The statute centers on four renewable fuel categories—conventional biofuel, advanced biofuel, cellulosic biofuel, and biomass-based diesel—each with its own target volume. A key part of the statutory definition of each fuel category is whether the fuel achieves certain greenhouse gas (GHG) reductions relative to gasoline and diesel fuel. Each fuel is assigned a lifecycle GHG emission threshold (in proportion to baseline lifecycle GHG emissions for gasoline and diesel).

The total renewable fuel requirement under the RFS is met with the combination of fuels from two renewable fuel categories: conventional biofuel and advanced biofuel. The requirement for advanced biofuel, in general, can be met with the combination of three types of advanced biofuel: cellulosic biofuel, biomass-based diesel, and other advanced biofuels. To date, the total annual volumes required have been met mostly with conventional biofuel (e.g., corn starch ethanol). Beginning in 2015, the mandate implicitly capped the conventional biofuel volume amounts while increasing the requirement for advanced biofuels. For instance, the statutory RFS total advanced biofuel requirement increases over time from approximately 7% of the total renewable fuel requirement in 2010 to 58% of the total renewable fuel target in 2022.

Differences Between the 2005 ("RFS 1") and the 2007 ("RFS 2") Laws

There are at least five major changes in the RFS as expanded in 2007 by EISA:

- Larger annual volume targets specified in statute for an extended period of time (i.e., through 2022).
- The establishment of separate requirements for different classes of biofuels (e.g., cellulosic, advanced).
- The addition of greenhouse gas accounting requirements.
- A different renewable biomass definition.
- An expansion of EPA's waiver authority to lower RFS volumes.

The renewable biomass definition for the RFS under EISA does not make the majority of woody biomass on federal lands available for use as a renewable feedstock. Further, the 2007 RFS waiver authority directs the EPA Administrator to set the annual standard for cellulosic biofuels under the RFS for the following year by November 30 of each year, and to lower the cellulosic biofuel standard if projected U.S. production is less than the volume in the statute. The 2007 RFS waiver authority also allows the EPA Administrator to reduce the renewable fuel and advanced biofuel requirements of the standard, if the cellulosic biofuel requirement is lowered.

Statutory Compliance

EPA regulates compliance with the RFS using a tradable credit system. Obligated parties (generally, refiners and importers) submit credits—called renewable identification numbers (RINs)—to EPA for each gallon in their annual obligation. (Thus, generally, each gallon of fuel produced to meet the obligation generates its own unique RIN.) In short, the annual obligation, referred to as the renewable volume obligation (RVO), is the obligated party's total gasoline and diesel sales multiplied by the annual renewable fuel percentage standards announced by EPA. The RVO is used by an obligated party to determine how many RINs they are to submit to EPA to be in compliance with the mandate.

In short, the RIN lifecycle can be described in three steps:

- A RIN is attached to a gallon of qualifying renewable fuel once that fuel is produced.
- The RIN is separated once the renewable fuel is blended with gasoline or diesel fuel.
- The separated RIN may be submitted for compliance, traded, or banked for future use.

RINs are valid for use to demonstrate compliance in the year they are generated and the following year. At times, an obligated party may carry a deficit due to a variety of reasons. If an obligated party is in a deficit, they are unable to submit enough RINs to meet their RVO for that compliance period. Obligated parties may carry a deficit from one year to the next, but in the year following the deficit, the obligated party must meet compliance for that year's renewable fuel volume requirement and purchase or generate enough credits to satisfy the deficit from the previous year. RINs may be used by the party that generates them or they may be traded with other parties. The EPA Moderated Transaction System (EMTS) is used to register RIN transactions. Different biofuels are not treated equally within the RFS. The categories are nested within each other, such that some fuels qualify for multiple categories (e.g., cellulosic ethanol), while others (mainly corn starch ethanol) may only be used to meet the overall RFS but not the advanced category or its nested subcategories. For example, a gallon of cellulosic biofuel may be used to meet the cellulosic biofuel mandate, the advanced biofuel mandate, or the total renewable fuel mandate, possibly making it a more highly valued fuel.

In addition, some biofuels generate more RINs per volume than others because of the difference in the fuel's energy content. This difference is accounted for by a metric referred to as the equivalence value (EV) of the biofuel. The EV of a renewable fuel represents the number of gallons that can be claimed for compliance purposes for every physical gallon of renewable fuel used, and it is generally the ratio of the energy content of a gallon of the fuel to a gallon of ethanol. For example, because biodiesel has an EV of 1.5 when being used as an advanced biofuel, 1,000 physical gallons of biodiesel would equal 1,500 RIN gallons of advanced biofuels.

The 2020 Final Rule

EPA announced the 2020 final rule for the RFS on December 19, 2019, and published the proposed rule in the Federal Register on February 6, 2020. The final rule calls for 20.09 billion gallons of total renewable fuel for 2020—an increase of 170 million gallons from the 19.92 billion gallons required in 2019. The implied conventional biofuel volume requirement remains at 15.00 billion gallons. The volume requirements finalized by EPA for 2020 for total renewable fuel, advanced biofuel, and cellulosic biofuel are all less than the volumes called for in statute but greater than the previous year's volumes—following a pattern that started in 2015 with one exception, the 2018 requirement for cellulosic biofuel. For 2020, EPA used the cellulosic waiver authority to reduce the statutory volumes for the total renewable fuel and advanced biofuel volume requirements "by the same magnitude as the reduction in the cellulosic biofuel reduction" (i.e., 9.91 billion gallons). EPA reports "the cellulosic waiver provision is best interpreted as requiring that the advanced biofuel and total renewable fuel volumes is reduced by equal amounts." EPA reports it found that the 2020 advanced biofuel statutory target of 15 billion gallons cannot be reached "primarily due to the expected continued shortfall in cellulosic biofuel; production of this fuel type has consistently fallen short of the statutory targets by 90 percent or more" EPA kept the biomass-based diesel standard for 2021 the same as for 2020—at 2.43 billion gallons.

EPA addresses other implementation duties in the 2020 final rule. For instance, EPA addresses a court decision that remanded the 2016 final rule back to EPA. EPA reports it is still actively considering the issue and will defer action on this issue which it anticipates happening in early 2020. Additionally, EPA changed the calculation for the annual percentage standards to account for volumes of gasoline and diesel projected to be exempt. Lastly, EPA finalized select regulatory changes put forth in both the 2020 proposed rule (e.g., clarification of diesel RVO calculations, clarification of EPA's authority to enforce pathway petition conditions, and revising the definition of an exporter of renewable fuel) and the 2016 Renewables Enhancement and Growth Support Rule (REGS Rule) (e.g., providing flexibilities for renewable fuel blending for military use, expanding the definition of heating oil to include heating oil used for cooling, and finalizing new pathways for co-processed cellulosic diesel, jet fuel and heating oil that are derived from co-processing biomass with petroleum to qualify as cellulosic biofuel).

The Blend Wall

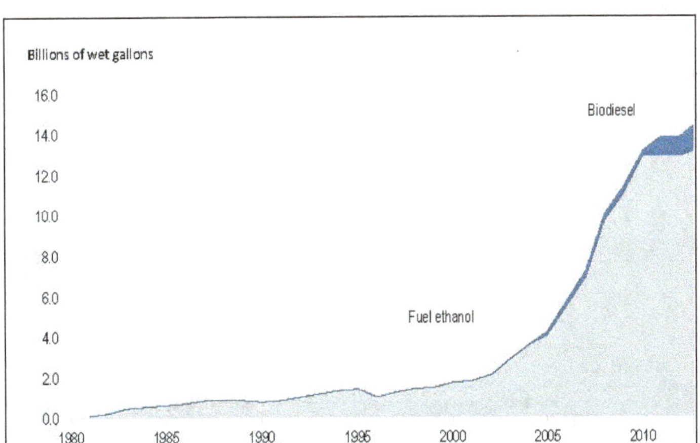

Fig: U.S. fuel ethanol and biodiesel consumption, 1981-2013.

Over the past ten years, the biofuel content of the US fuel supply has risen from less than 2 billion gallons in 2000 to more than 14 billion gallons in 2013. As figure shows, most of that growth has been in ethanol (primarily corn ethanol and some cane ethanol). The sharp increase in ethanol consumption over the past decade is a result of several factors, including the phasing out of MTBE as oxygenate and its replacement with ethanol as well as state and federal biofuels policy (including the ethanol blenders' tax credit and the RFS).

US ethanol prices have historically moved with gasoline prices. As is shown in Figure below, since 2007 the wholesale price of corn ethanol has typically been below the price of wholesale petroleum gasoline (RBOB) on a volumetric basis, but above the petroleum gasoline price after adjusting for ethanol having only 68% the energy content of petroleum gasoline per gallon. Making precise inferences from these data either about the effect on retail gasoline prices of blending ethanol or on the underlying actual cost of production of ethanol faces two difficulties. First, because ethanol is used to boost octane and replaces petroleum octane boosters, it is competing both with gasoline on an energy basis and with the octane boosters, enhancing the value of ethanol. Second, the price of traded ethanol is influenced by a host of subsidies and policies, complicating the relationship between the production cost of ethanol and the wholesale traded price of ethanol. During the period after the expiration of the volumetric ethanol excise tax credit on December 31, 2012, and before high RIN prices in February 2013— when ethanol was approximately 10% of gasoline and was not receiving a direct subsidy from either the tax credit or the RFS—ethanol prices averaged approximately 20% less than wholesale gasoline on a volumetric basis, and approximately 15% more on an energy-adjusted basis. This episode happened to coincide with the drought of 2012, and during the first six months of 2012, before the severity of the drought became clear later in the summer, the price of ethanol was nearly 30% less than petroleum gasoline on a volumetric basis, and 5% above on an energy-adjusted basis. Ethanol blending decreased US gasoline prices by up to $0.10 per gallon. Since the summer of 2014, ethanol prices have fallen with gasoline prices.

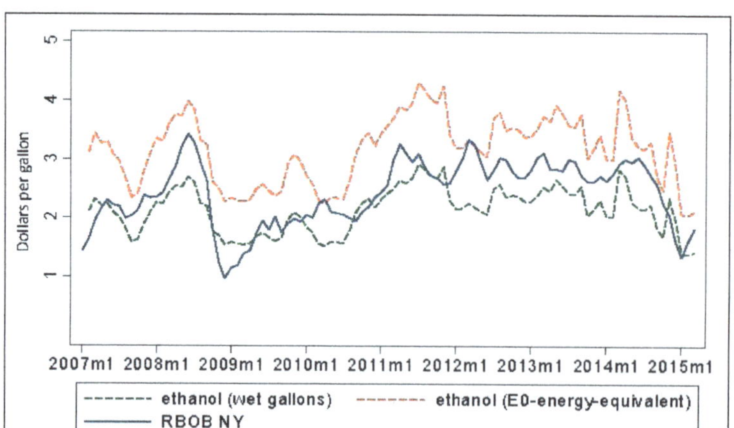

Fig: Price of wholesale gasoline and ethanol on an energy-equivalent basis.

As can be seen in figure above, the composition of biofuel consumption has changed since 2010. These changes are largely a consequence of the E10 blend wall, the RFS, and various biofuels tax credits. The EISA statutory volumes hit the E10 blend wall several years earlier than expected based on gasoline consumption projections at the time the EISA was developed and passed. As Figure 4 shows, the projected US gasoline consumption to grow into the indefinite future. But as a result of the Great Recession, new vehicle fuel economy standards, high gasoline prices, and possible changes in driving habits, total gasoline consumption has fallen, not increased, and the EIA's current estimate of 2014 gasoline consumption (including blended ethanol) is 137 billion gallons (Bgal), 15% below the 161 Bgal in the 2014 reference scenario projection in the EIA's 2006 Annual Energy Outlook. Based on the 2006 EIA projection, the 2014 E10 blend wall would be at approximately 16.1 Bgal of ethanol, whereas based on current estimates, it is at 13.7 Bgal, 2.4 Bgal less than based on the 2006 forecast.

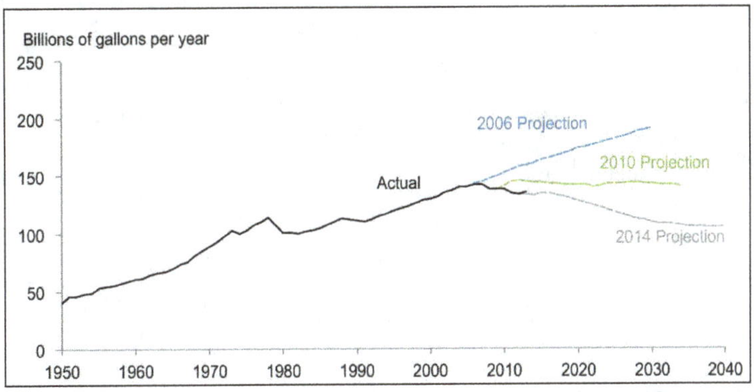

Fig: U.S. Consumption of Motor Gasoline, 1950-2040.

If the conventional RVO exceeds 10% of gasoline consumption, then the conventional RVO cannot be filled by corn ethanol blended into E10 alone. This is the situation commonly referred to as the RFS mandate exceeding the E10 blend wall, although as discussed below this is more accurately not a "wall" but rather a situation in which additional ethanol must be provided through higher blends. Through 2012, production and consumption of corn ethanol exceeded the RFS conventional mandate, that is, the conventional mandate did not bind. Figure below shows the relation between implied mandated ethanol consumption (shown as a range that adds statutory cellulosic volumes to the statutory conventional RVO), actual ethanol consumption, and 10% of actual gasoline consumption. In 2013, the conventional RVO of 13.8 billion gallons constituted 10.3% of gasoline consumption, exceeding the E10 blend wall. Because of very low penetration of higher blends, the resulting conventional biofuels shortfall was met through a combination of RINs banked from consumption in excess of the RVO in previous years, D4 and D5 RINs that were produced in excess of their respective RVOs, and D6 RINs generated by non-ethanol conventional fuels. In 2014 the statutory conventional RVO increases to 14.4 Bgal and, as shown in table below, it reaches its cap of 15 Bgal in 2015. That 15 Bgal cap is nearly 11% of currently projected gasoline consumption and exceeds the ethanol capacity of E10 by roughly 1.3 Bgal.

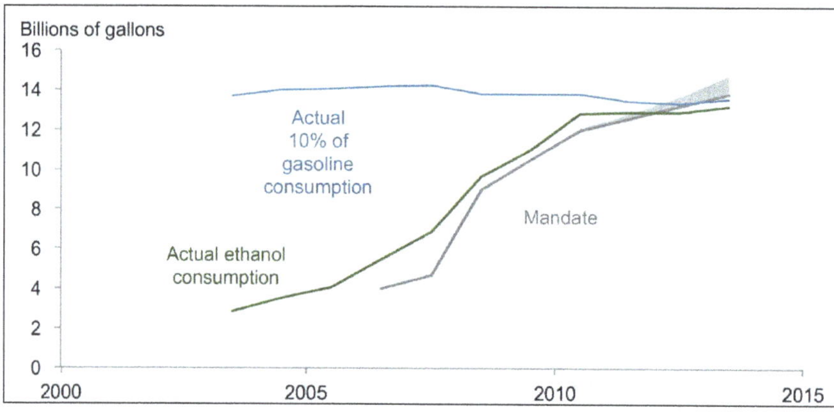

Fig: U.S. ethanol and gasoline consumption.

Year	Statutory Volumes (Billions of gallons)					As Set by EPA rule-making (billions of gallons)				
	Cel-lu-losic	Bio-mass-based Diesel	Total ad-vanced	Total renew-able	Im-plied con-ven-tional pool	Cellu-losic	Bio-mass-based Diesel	Total ad-vanced	Total renew-able	Implied con-ven-tional pool
2009	n/a	0.5	0.6	11.1	10.5	n/a	0.5	0.6	11.1	10.5
2010	0.1	0.65	0.95	12.95	12	0.00065	0.65	0.96	12.95	12
2011	0.25	0.8	1.35	13.95	12.6	0.006	0.8	1.35	13.95	12.6
2012	0.5	1	2	15.2	13.2	0.0105	1	2	15.2	13.2
2013	1	≥1.0	2.75	16.55	13.8	0.0008	1.28	2.75	16.55	13.8
2014	1.75	≥1.0	3.75	18.15	14.4					
2015	3	≥1.0	5.5	20.5	15					
2016	4.25	≥1.0	7.25	22.25	15					
2017	5.5	≥1.0	9	24	15					
2018	7	≥1.0	11	26	15					
2019	8.5	≥1.0	13	28	15					
2020	10.5	≥1.0	15	30	15					
2021	13.5	≥1.0	18	33	15					
2022	16	≥1.0	21	36	15					
2023	b	b	b	b	b					

Economics of the RFS

The RFS provides a guarantee to biofuel producers that they will be able to sell up to the mandated volume for a given year. If the biofuel can be produced for less than the price of its petroleum alternative and if there are no non-price market failures or other impediments to the consumption of renewable fuels, then that fuel will enter the fuel supply for price reasons, not because it is required to by the RFS. If, however, the marginal cost of producing the biofuel exceeds what consumers are willing to pay, then a subsidy is needed to make sure the fuel is produced and consumed. Because RINs are separated by blending a biofuel into the fuel supply, and RINs must be turned in to the EPA when an obligated party (a refiner or importer) sells petroleum fuel into the fuel supply, the price of a RIN is the vehicle for transferring corrective production subsidies to ensure that biofuels are produced and consumed at the mandated level. By the same token, the price of RINs is a measure of the compliance cost of the program: the greater is the RIN price, the greater is the value of the RINs that the obligated party must turn in. For these reasons, understanding the theory and empirical behavior of RIN prices is central to understanding the economics of the RFS.

It begins with a discussion of the fundamental determinants of RIN prices in both static and dynamic (bankable) settings. It then turns to the effect of RIN prices on final transportation fuels, both in theory and empirically. In theory, the cost of RINs should be passed through to consumers, increasing the price of fuels with low renewable content (like diesel, which on average contains roughly 3% renewables) and decreasing the price of fuels with high renewable content (like E85). Consistent with the theory, empirical evidence indicates that the price of diesel and petroleum gasoline (E0) rises with RIN prices, and the price of E10 does not vary with RIN prices. Theory also predicts that the price of E85 should fall when RIN prices rise, but the evidence suggests that there is incomplete pass-through of the RIN price subsidy to retail E85 prices, so that only part of the effective RIN subsidy for E85 is passed along to the consumer.

RIN Price Determination

The price of RINs, like other assets, depends on underlying fundamentals. For RINs, the fundamental is the difference between the price necessary to produce and distribute the mandated quantity of the relevant biofuel and the price the consumer is willing to pay for it. Because of the nested fuel structure of the RFS, RIN prices can further depend not just on market conditions for the fuel generating the RIN, but on markets for other biofuels. In addition, because the RIN is bankable, the price of a RIN today depends on expected future fundamental values as well as the fundamental values today. Because the fundamental values depend on the RFS mandated volume, the price of the RIN today depends on current RFS policy and on expected future RFS policy.

RIN Price Fundamentals and Annual Subsidy Values

The fundamental factor in determining the RIN price is the value of the subsidy needed in a given compliance year to produce and consume the relevant biofuel at the mandated level. For the moment, suppose that RINs must be used in the year they are generated, and ignore interactions between fuels induced by the RFS nesting. The basic idea of the RIN price fundamental is that a biofuel producer receives two payments when she blends a gallon of biofuel: one for the physical product that the consumer uses as fuel, the other for the RIN that is generated and sold when the biofuel is blended. In equilibrium, the price at which the producer is willing to sell the RIN will just cover the marginal cost of producing the required volume. Thus the annual subsidy value, or RIN price fundamental, is determined by the difference between the price producers require for the marginal gallon of biofuel and the price consumers are willing to pay for it.

This RIN price fundamental is illustrated for biodiesel in figure below. The price that a producer requires to produce the mandated volume is the price on the biodiesel supply curve (the "supply price") at that volume. This supply curve is upward-sloping because as more biodiesel is produced, feedstock costs (and perhaps other marginal production costs) go up. Similarly, the price that consumers are willing to pay is the price on the demand curve (the "demand price") at that volume. In the figure, this demand curve is flat at the price of petroleum diesel, so that at current volumes, biodiesel is in effect interchangeable with petroleum diesel. In the figure, the supply curve lies above the demand curve because biodiesel is more expensive than petroleum diesel. The difference between these two prices is the amount that must be covered by selling the RIN. Thus, the RIN price fundamental is the difference between the supply price and demand price as shown in Figure below, that is, the annual subsidy value. In the case illustrated in the figure, the subsidy value flows to the biodiesel producer because consumers are indifferent between using diesel and biodiesel, but producers cannot afford to produce volumes greater than the market equilibrium Q_o without a subsidy.

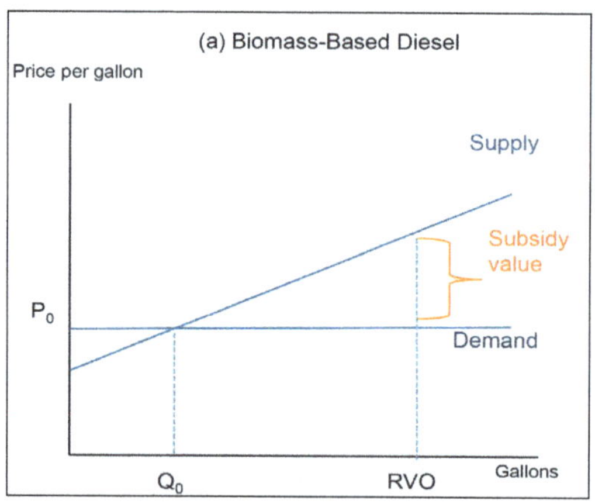

Fig: Biomass-based Diesel.

Figure below provides a qualitative illustration of the corn ethanol market, in which the nonlinear demand curve for ethanol is a stylized representation of the blend wall. Below the equilibrium level Q_o, demand is insensitive to price as long as the price of ethanol is less than the price of petroleum gasoline. But for volumes exceeding Q_o, it becomes increasingly difficult to put additional ethanol into the market, so that a sharply growing consumer subsidy is needed. In this sense, the blend wall is not so much a "wall" but an inflection point after which sharply increasing subsidies are need to ensure consumption of incremental gallons of ethanol. In the case illustrated—in which the RVO is in the blend wall portion of the demand curve—most of this subsidy flows to the consumer in the form of lower ethanol prices, since not much of a supply price increase is needed to induce the small additional amount of ethanol production.

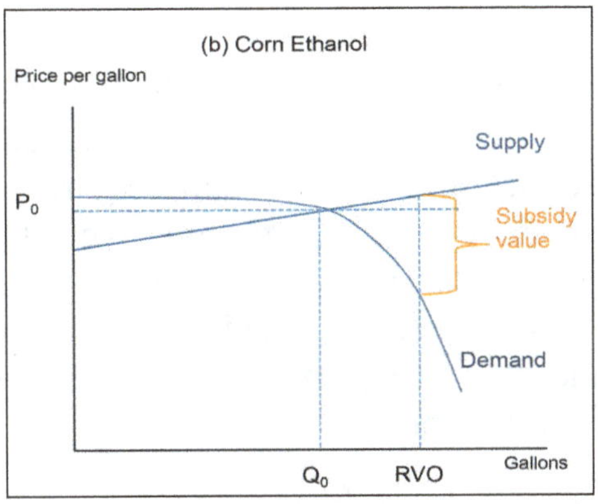

Fig: Corn ethanol.

RIN Price Implications of the RFS Nesting Structure

Under the nested fuel structure of the RFS shown in figure above, D4 RINs can be used to satisfy the biomass-based diesel requirement, the total advanced requirement, or the total renewable requirement; D5 RINs can be used to satisfy either the total advanced or total renewable requirement; and D6 RINs can only be used to satisfy the total renewable requirement. This nesting structure implies that a D4 RIN is at least as valuable as a D5 RIN, and a D5 RIN is at least as valuable as a D6 RIN. Moreover, as a result of this nesting structure there are different RIN price "regimes," depending on which of the different requirements are binding, that is, which fuels if any are produced in excess of their requirement to generate RINs to satisfy an obligation within which it is nested.

To make this concrete, consider the example of figure above in which there are only two fuels, biodiesel and corn ethanol, and two RINs, D4 and D6. If the biodiesel subsidy value is less than the ethanol subsidy value, then biodiesel producers can produce in excess of the biodiesel requirement and sell the resulting D4 RINs for the purpose of satisfying the total renewable (conventional fuel) mandate. In this scenario, the amount of biodiesel produced will rise, and the amount of conventional ethanol will fall, to the point that the subsidy values in the two markets are equalized, so the D4 and D6 RINs have the same price. In contrast, if the subsidy value for biodiesel at the biodiesel requirement exceeds the subsidy value for conventional ethanol at its requirement, then biodiesel producers will have no incentive to produce excess biodiesel and the price of the D4 RIN will exceed the price of the D6 RIN. More generally, the nesting structure implies the price inequalities, PD4 ≥ PD5 ≥ PD6 ≥ 0 (where PD4 is the price of the D4 RIN, etc.). If the inequality is strict, then the mandate is binding, for example if PD4 > PD5 then no biodiesel is being produced in excess of the biomass-based diesel mandate. In contrast, if PD4 = PD5 = PD6, then both biodiesel and non-diesel, non-cellulosic advanced fuels are being produced in excess of their mandates, and one or the other or both fuels are being used to generate RINs to satisfy the total renewable obligation.

Bankability Implies that Current RIN Prices Incorporate Future Subsidy Values and Policy Expectations

Suppose that the requirement is low this year but is expected to rise next year, so the subsidy value needed to meet the requirements will rise. Because RINs are bankable, their price will rise above this year's subsidy value in anticipation of next year's stiffer requirement, inducing production in excess of the requirement this year. The excess RINs are banked and used next year. In this example, bankability increases RIN prices this year and lowers them next year. Because RINs can, in effect, be rolled over indefinitely (subject to the 20% rollover cap), this logic further extends to future years. Thus RIN prices today reflect market participants' views about the stream of subsidy values extending for the life of the program.

RINs have features that are similar to financial options. First, they can be exercised (retired) for compliance this year or in the future. Second, they have a nonlinear payoff function: the fundamental value of the RINs is positive if the annual subsidy value is positive, otherwise it is zero (because RINs need never be retired, their price cannot be negative). This latter feature is analogous to a financial option having exercise value only if it is in the money.

Like a financial option, the price of RINs today will increase if there is an increase in uncertainty about future fundamental (subsidy) values. This uncertainty arises from uncertainty both about future economic fundamentals (weather, driving demand, oil prices, and geopolitical factors) and about future policy. For example, suppose a multi-year policy path is announced with low RVOs. Because legal uncertainty and the annual rulemaking process make it difficult to commit convincingly to such a path, there will remain a chance that policy might return to the more expansive path of the statutory RVOs, and as a result there is a chance that the annual subsidy value will be high in the future. This possibility of high future RIN prices in turn adds in a premium (which could variously be called a risk premium, or time value, or option value) arising solely from uncertainty. Thus RIN prices depend on both, expectations of future policy and on the degree of certainty about future policy.

RIN Prices Since 2012

Figure below shows the daily price of D4 (BBD), D5 (advanced), and D6 (conventional) RINs from July 2012 through March 24, 2015. Through the end of 2012, D6 RIN prices were low (less than $0.10) and the three RINs had distinct prices, indicating that the BBD, total advanced, and total renewable mandates were each binding. With increasing market awareness of the blend wall and with the release of the 2013 proposed rule, D6 RIN prices rose from mid-January through February 2013, and fluctuated around $0.75 from March through mid-May 2013. At this point, D6 RIN prices exceeded the 2012 D5 and D4 RIN prices, the BBD and Total Advanced RVOs became nonbinding—incentives were created to produce more advanced renewables—and excess D4 and D5 RINs were generated for compliance with the D6 mandate. Indeed, for most of the period since February 2013, the three RIN prices have moved in tandem, with the D4-D5 and D5- D6 spreads averaging \$0.03 and \$0.06, respectively, from March 2013 through November 30, 2014. These spreads arguably represent the greater option value associated with the lower RIN numbers: the option value for a D4 biomass-based diesel RIN will exceed that for a D5 RIN if there is some probability that the BBD RVO will be binding in the future (at which point the biomass-based diesel subsidy value will rise above the advanced fuel subsidy value). RIN prices separated again in early 2015 as petroleum prices fell.

Even though supply and demand conditions were relatively stable through 2013, with crop production recovering from the 2012 drought and the US and global economic growth relatively stable, 2013 saw large fluctuations in RIN prices. The initial run-up in RIN prices in early 2013 was due in part to the increasing awareness of the E10 blend wall, and some of the short term volatility could have been due to the markets being thin. The major source of the fluctuations, however, was arguably changing market expectations about the future course of policy. RIN prices rose after the release of the 2013 proposed rule, which acknowledged the blend wall but indicated that there would be no adjustments to the total renewable or total advanced statutory RVOs, and that additional gaps between the RVO and the blend wall could be met by drawing down RIN stocks.

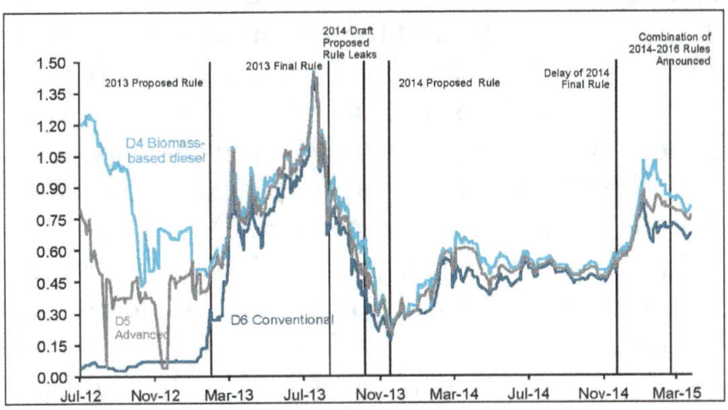

Fig: Daily RIN prices, July 1, 2012 - March 24, 2015.

RIN prices fell substantially after the release of the 2013 Final Rule, which gave forward guidance indicating that, unlike the 2013 rule, the 2014 rule would be set bearing in mind the constraints of the blend wall. RIN prices fell further upon the leaking of a draft of the 2014 proposed rule, which proposed to use both the cellulosic and general waiver authorities to implement this guidance by backing the RVOs out of estimates of the total amount of ethanol that could be introduced into the fuel supply; the 2014 proposed rule implied 10.09% ethanol content, just over the E10 blend wall after taking into account a small amount of E85 sales. RIN prices rose subsequently based on evolving perceptions of the various pressures facing the EPA, including public statements that the RVOs could increase from the proposed to final 2014 rule and, most recently, the announcement that the 2014 RVOs would be finalized in 2015.

Effects of RIN Prices on Fuel Prices

A central, highly charged question surrounding the RFS is the effect of RIN prices on pump prices.33 In brief; RINs act as a tax on fuels with low renewable content and a subsidy to fuels with a high renewable content in the long run (in equilibrium) the RIN price serves both to increase the supply of biofuels, relative to petroleum, and to increase consumption, and a RIN price increase is passed along in part to producers, who produce more, and in part to consumers. Because much of the debate has focused on the short-run link between RIN prices and fuel prices, the discussion here focuses on the short run, over which supply does not change and RIN prices change not because of current supply and demand considerations but for some other reason, such as changes in policy expectations.

To be concrete, consider RIN obligations in 2013, when the required RIN bundle consisted of a total of 0.0974 RINs (the sum of the required D3, D4, D5, and D6 RINs). Suppose all RIN prices increase by $1, so that the price of a RIN bundle increases by $0.0974. Because E10 is 90% petroleum, selling 0.9 gallons of petroleum would increase RIN costs to the obligated party by the cost of 0.9 RIN bundles, that is, by 0.9 x $0.0974 = $0.088. But blending 0.1 gallon of ethanol into E10 generates 0.1 D6 RIN, which the blender can sell for $0.10. In a competitive market, the petroleum producer passes on the $0.088 extra cost, the blender (who gets to sell the RIN) passes on the $0.10 savings, and the consumer comes out ahead by $0.012. Repeating the calculations above for diesel (which has a low renewable content) indicates that a $1 increase in the price of all RINs results in approximately a $0.05 increase in the price of diesel per gallon under perfect competition. Repeating the calculations again for E85 (which has on average 74% ethanol) results in a predicted decrease in the E85 price of $0.72 per gallon. Thus, with competitive markets and complete RIN price pass-through, if all RIN prices are $1, the net effect is a small subsidy to E10, a large subsidy to E85, and a tax on diesel, which has the lowest renewable content. The specific values of the tax and subsidy depend on RIN prices. For example, in mid-February 2015, D6, D5, and D4 RIN prices were approximately $0.70, $0.80, and $0.85, respectively; at those prices, with perfect pass-through their theoretical effect was to increase the pump diesel price by $0.03 per gallon, to decrease the E10 price by $0.01, and to decrease the E85 price by $0.50. Although specific tax and subsidy values depend on the RIN prices and the obligation percentages, the structure has the effect of taxing the lowest-renewable final fuel (diesel) and subsidizing the highest-renewable fuel (E85), with E10 receiving a slight subsidy because it has slightly more renewable content than the 2013 total renewable fractional obligation.

Figure below provides empirical evidence on two of these short-run pass-through predictions, for wholesale gasoline and for E10. On the margin, a refiner could choose to sell a gallon of gasoline into the US market, where the refiner would incur the RIN bundle cost, or export it.

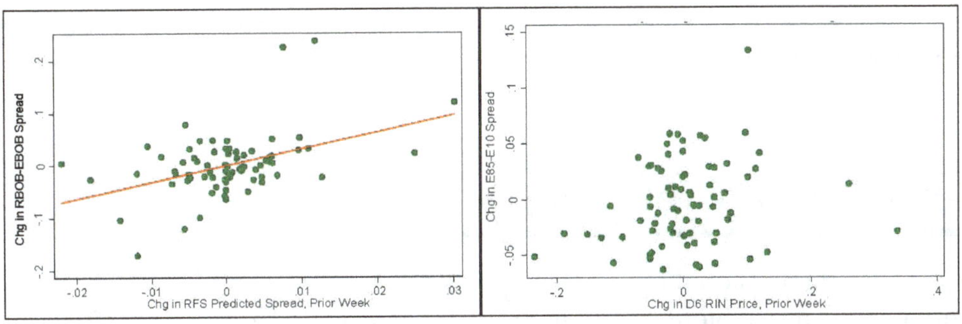

As such, in equilibrium, the wholesale price of gasoline in the United States (RBOB) should equal the international price, plus the per-gallon price of the RIN bundle, differential transportation costs, and other tax and fee differentials. To the extent that transportation costs and other taxes and fees either do not change, or have changes that are unrelated to changes in RIN prices, exogenous changes in the price of the RIN bundle should translate one-for-one into changes in the spread between RBOB and international wholesale gasoline prices.

This pass-through prediction is examined in figure 8a, which plots the weekly change in the spread between RBOB and EBOB versus the previous week's change in the price of the RIN price bundle. The scatterplot shows that, on average, changes in RIN prices in the previous week are positively associated with changes in the RBOB-EBOB spread. Once lags are taken into account, the empirically estimated pass-through is consistent statistically with complete pass-through, which is to say, US wholesale prices generally rise when RIN costs increase, and fall when they decline.

Figure examines the relation between changes in the E10 pump price changes in the D6 RIN price in the previous week. Consistent with the theory outlined above, there is negligible estimated effect of RIN prices on pump E10 prices.

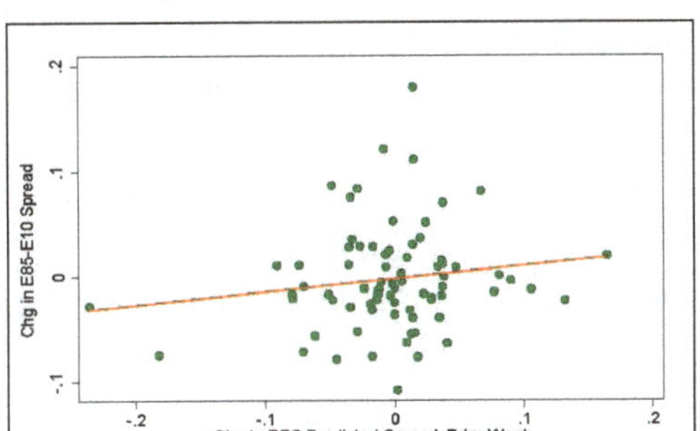

Fig: E85-E10 spread v. prior week RFS-predicted.

Figure shows the relation between the change in the spread between E85 and E10 average retail prices and the predicted change in the spread based from previous week changes in RIN prices. Consistent with theory, the relationship is positive (when the E85 price is predicted to drop, relative to E10, it does on average); however, the estimated pass-through is less than one-for-one. Regression analysis that includes lagged effects suggests that of a $1 increase in RIN prices, roughly one-third is passed through to consumers in the form of lower E85 pump prices.

Biofuel Policy in Europe

Biofuels were first promoted at an EU level with the introduction of The Directive on the promotion of the use of biofuels or other renewable fuels for transport. The Directive set indicative targets for the percentage share biofuels should achieve in the total volume of fuel consumed in EU transport (2% by 2005 and 5.75% by 2010). Member States were free to determine how they would meet the targets and were required to submit annual reports indicating the progress made towards reaching them. The Energy Taxation Directive, passed in the same year, allowed tax exemptions and tax reductions under certain conditions to permit Member States to favour biofuels in this way if desired.

Although the targets were only indicative, biofuel use more than doubled between 2003 and 2005. Despite this mini boom, by 2005 the share of biofuels in the EU's transportation fuel total had reached only 1.4%, falling short of the 2% target.

In February 2006, the European Commission released the communication "An EU Strategy for Biofuels" 34 final) based on the Biomass Action Plan. This strategy set out seven strategic policy areas to facilitate the development of biofuels in both the EU and developing countries. These areas included support to research and development; support to developing countries; a study into trade and custom opportunities available to support biofuels; and available mechanisms to increase demand for biofuels. Under the last heading, the Commission stressed the importance of national targets, use obligations and ensuring the sustainable production of biofuels.

In 2007 the Renewable Energy Roadmap proposed a legally binding target of 20% for renewables as a contribution to total energy consumption by 2020, with a minimum target for biofuels of at least 10% of overall consumption of petrol and diesel in transport. In March 2007 this recommendation was endorsed as EU leaders committed to transform Europe into a highly energy-efficient and low-carbon economy. They pledged to cut CO_2 emissions to 30% below 1990 levels by 2020, upon the agreement of a comprehensive global accord. The EU also committed to reduce its emissions to 20% below 1990 levels regardless of a global agreement. These emission targets included three key energy objectives:

- A 20% reduction in energy consumption through improved energy efficiency by 2020.
- A 20% share of renewable energy in gross energy consumption by 2020.
- A 10% share of consumption of petrol and diesel in transport to be met by biofuels by 2020.

In January 2008 the European Commission proposed a package of legislative measures to implement those targets pledged in March 2007. Following negotiations the binding measures were agreed by EU leaders and the European Parliament in December 2008. The package, which came to be known as the "Climate and Energy Package", was adopted in April 2009. Its goals, which became known as the "20-20-20 targets", are intended to ensure that the EU meets its pledges made at Kyoto and to set the stage for further global negotiations.

During the negotiation procedure in 2008 many of the Commission's proposals were altered to address the concerns of various stakeholders and Member States. Although the overall targets remained the same many of the details were changed, and in some cases weakened. The main change to the proposal with regards to biofuels was the widening of the remit of the 10% target. The target, which originally stated that 10% of transport fuel should be met by biofuels, was changed to state that 10% of the energy in transport must be sourced from renewable resources. This change in wording allows electricity and hydrogen to be included in the target, in the hope that it will act as an incentive for developing these technologies.

Although policy is mandated at a continental level by the European Union, Member States are largely free to determine how best to implement the policy frameworks and meet the targets set. Table below shows consumption levels of biofuels in countries with the highest use in 2008.

Table: Highest consumption levels of biofuels in the EU by country in 2008.

Country	Bioethanol	Biodiesel	Total Consumption
Germany	4020000	2477983	3257186
France	403510	2020690	2424200
United Kingdom	105189	691335	796524
Spain	125000	519000	644000
Italy	-	557280	557280
Poland	213968	340560	459354

Main Mechanisms used to Support Biofuels Within the EU

The promotion of biofuels in the EU has been implemented using various policy instruments. Several directives have dealt with differing aspects of the biofuels production chain. The Energy Taxation Directive, for example, allowed Member States to exempt biofuels from being taxed; the Fuel Quality Directives allowed blending and the Common Agricultural Policy supported the growth of energy crops. The list below gives the main tools which are/have been used to promote biofuels in the EU:

- Renewable Energy Directive (RES-D): Directive.
- Fuel Quality Directive (FQD): Directive.

- EU Climate and Energy Package: Text adopted in the sitting of 17th December 2008.
- Biofuels Directive: Biofuels Directive 2003/30/EC.
- Fuel Quality: Directive II Directive 2003/17/EC.
- Fuel Quality: Directive I Directive.
- Energy Taxation Directive: Directive 2003/96/EC.
- Common Agricultural Policy.

The Energy Taxation Directive passed in 2003 allowed Member States to apply, for the first time, differentiating tax rates to biofuels or to exempt them from being taxed. This enabled Member States to offer support to biofuels by reducing the cost difference between fossil fuels and biofuels. Member States applied the tax exemption scheme in a variety of ways to differing degrees. The success of its implementation also varied widely between states, with some Member States achieving high levels of penetration, whilst others were less successful.

Despite the variable success, partial or total tax exemptions for biofuels have been vital in promoting biofuels in the EU. All Member States who have succeeded in achieving high biofuel penetration rates have, or have had, differentiating tax schemes in place. Tax exemption schemes have been most successful in States with high fossil-fuel tax rates. This is due to the fact that, for biofuels to be economically competitive with fossil fuels, a tax exemption must be applied that is large enough to cover their higher production costs. For tax exemption schemes to work well, the fossil fuel tax must be high enough to result in fossil fuels price which is higher (after tax) than the price of biofuels (with or without partial tax).

Whilst such schemes have been successful in introducing and increasing the levels of biofuels, their application has resulted in significant revenue losses for governments. Consequently, a trend of Member States switching from tax exemption schemes to obligation schemes can currently be observed. Obligation schemes prescribe the mandatory uptake of biofuels in the market. Here, fuel suppliers are required to achieve a certain share of biofuels in their total sales. In such schemes, the additional costs are carried by the fuel suppliers and ultimately by the transport users instead of being met by the government. Since 2005 twelve EU Member States, accounting for almost 90% of biofuel consumption in the EU, have switched from tax exemption schemes to obligation schemes. Although this mechanism has previously been a key tool used by the EU to promote biofuels, the current shift towards obligation schemes signals not only that its importance is likely to decrease in the future, but also the increasing unwillingness of the EU and its Member States to bear the cost of environmental policy.

The Fuel Quality Directives have also played an important role in increasing the market penetration of biofuels. They established fuel specifications for petrol and diesel and made allowances for the incorporation of biofuels. Blending limits were also set for biodiesel and ethanol. The most recent change to the Fuel Quality Directive was its revision, adopted in December 2008. The new Fuel Quality Directive amends Directive (98/70/EC) and enables higher blends of biofuels in petrol and diesel, such as the establishment of a new petrol blend allowing up to 10% ethanol. It also contains an obligation for fuel suppliers to reduce the lifecycle emissions of their fuels by 6% by 2020 compared to 2010 fossil fuel emission levels. The obligation is likely to promote the use of biofuels as it will encourage suppliers to use increasing levels of biofuels in their fuel mix due to their superior greenhouse gas emission performance.

Biofuels in Europe have also, until recently, been supported through agricultural policies. Since 1998 support has been available to domestic producers of biofuel feedstock. This support, established within the Common Agricultural Policy (CAP), was available through two instruments. Firstly, the Energy Crop Scheme which began in 2004 provided a direct subsidy of up to 45€/ha for farmers within the EU. This direct subsidy was available for a defined maximum area, which was finally reached in 2007. Once this limit was reached, however, virtually no energy crops were grown outside the scheme, i.e. without the subsidy support. Secondly, energy crops could also be grown on Set-aside land. In the EU, set-aside accounted for 10% of a farmer's land and could not be used for production under EU agricultural law. The CAP reform in 1992 allowed the production of non-food crops on set-aside land without losing the subsidy. In the Common Agricultural Policy Health Check in 2008 the EU agreed to abolish both the Energy Crop Scheme and the compulsory Set-Aside Scheme. The abolishment of compulsory set-aside came into force shortly afterwards and the Energy Crop Scheme will terminate in 2010. Consequently, by the end of 2010, there will be no remaining mechanisms to directly encourage the production of biomass feed stock in the EU.

The effect that the abolishment of these support mechanisms will have on the cultivation of biofuel feedstock in the EU is currently unknown. Biomass production and agricultural policy are inextricably linked and it is widely recognised that any changes in agricultural policy are likely to have significant effects on the biomass stock produced and thus the cost and the amount of biofuels available. The annulment of the support for biofuels previously provided through CAP signals a future demise in the importance of agricultural policy for biofuels. The consequences of these changes are not yet fully understood, although it may have been a telling sign in 2007, when the maximum area for the energy crop scheme was reached, that virtually no further energy crops were grown without its support. On the other hand it is assumed that strong demand created by the mandatory targets should enable farmers to economically produce biomass stock without financial support.

The previously described mechanisms have contributed significantly towards the development of biofuels during the last ten years. The recent Climate and Energy Package, however, nullifies or updates much of the previous legislation. Its implementation will have a profound impact on how biofuels are used and the level of market penetration achieved in the future. Within the package, the Renewables Directive has arguably the highest significance with regards to biofuels. The Directive deals with biofuels in several ways, the most noteworthy being the mandatory target which states that 10% of final energy consumption in transport should be met by renewable energy by 2020. Another important aspect of the Directive is the mandatory sustainability criteria to which all biofuels are subject. This aspect in particular has received high publicity and its detailing in the Directive has left serious questions open regarding indirect land use change and potential clashes with trading laws.

The 10% Target

The transport sector is vital to every nation's economy and that of the EU is no exception. What is perhaps notable about the EU's transport sector is the extent to which it is dependent on fuel imports. The EU's transport sector imports 98% of its fuel, 80% of which is sourced from Russia, Saudi Arabia, Libya, Iran and Norway. High import rates, combined with imports sourced predominantly from politically unstable countries exposes the EUs' transport sector to risk both politically and economically. This results in a sector which is highly vulnerable to both political and economic variations. The transport sector is also a major emitter of greenhouse gasses, contributing to 28% of the EU's total emissions.

Mandatory targets are deemed necessary by the EU to create a secure investing climate and to ensure its climate change goals are reached. Although the EU set a standard target for renewable energy consumption (20% of total energy consumption must be derived from renewables), it was felt that an explicit target was required for the transport sector, due to its unique features. Firstly, oil dependence in the transport sectors is one of the most serious issues of energy insecurity faced by the EU. Secondly biofuels and renewable energy in the transport sector in general, have higher production costs compared to other forms of renewable energy. There is a risk therefore that, if no specified requirement existed, other cheaper forms of renewable energy may be used that are applicable in other sectors but do not tackle the fuel insecurity or the rising emissions of the transport sector.

Whilst perhaps the best justified of the Climate and Energy Package targets, the 10% target has received the most criticism. Heightened public attention and heated discussion reached its peak during the summer of 2008 when multiple parties urged the EU to abolish its targets on a back drop of soaring food prices and issues of sustainability.

The Gallagher Report written by a panel of UK government experts, called on the EU to completely drop its mandatory targets and to substantially reduce its indicative targets. It also claimed that there was no proof that the targets could or would be sustainable and that they would have an impact on food prices, particularly in the developing world. Other international organisations have also published reports raising a variety of concerns, including the International Energy Agency (IEA), the Food and Agricultural Organisation of the UN (FAO), the Organisation for Economic Co-operation and Development (OECD) and the European Environmental Agency (EEA). These reports also raised issues such as food security, sustainability of production and commodity prices among others and most called for the EU to drop its targets.

Despite the criticism and the large debate, which was also heated within the EU, the target survived and remained in the final Directive. The Directive, which was adopted in April 2009, must be met by all Member States by 2020.

Sustainability Criteria

A large proportion of the criticism surrounding the 10% target focused on the sustainability of biofuels and their potential unintended social effects, particularly on food prices.

The EU, since first announcing its intention to set a mandatory biofuels target, has maintained that any production or use of biofuels must be sustainable. The Renewables Directive (2009/28/EC) aims to ensure this ambition is met through the use of mandatory sustainability criteria. The criteria set out three main requirements which biofuels must meet in order to be counted towards the target, or to be eligible to receive tax rebates or subsidies:

- The greenhouse gas emission savings from the use of biofuels and bioliquids must be at least 35% (rising to 50% in 2017) compared to fossil fuels.
- The feedstock of the biofuel is not to be derived from land with high biodiversity value such as highly biodiverse grassland.
- The feedstock of the biofuel is not to be derived from land with a high carbon stock.

These criteria apply to biofuels and bioliquids and for both imported and domestically produced feedstock.

The sustainability criteria laid out by the Directive focus on ensuring that biofuels give greenhouse gas emission savings and on protecting biodiversity. The Directive sets out a method for calculating life cycle greenhouse gas emissions and includes a list of default values for savings achieved using different biofuel production pathways. CO_2 emissions resulting from direct land conversion are to be counted in the life cycle analysis of biofuels.

Although the Directive aims to ensure the sustainability of biofuels through the use of sustainability criteria, many feel that those included in the Directive are weak and leave many loop holes open. The problem lies in the detail; in particular how the Directive defines the land type which is to be protected. Highly biodiverse and high carbon stock areas are defined so that many such areas fall outside the scope of the Directives definition and thus outside of its protection. For example the EU rejected the internationally recognised forest definition, endorsed by the FAO among others, in favour of its own definition of a 30% canopy cover. This variation in definition could potentially result in up to 50% of global forested areas falling outside the protection of the sustainability criteria and thus being open to conversion for biofuels.

Another strong criticism against the sustainability criteria used in the Directive is their inability to deal with the social effects of increased biofuel production including; food insecurity, volatile commodity prices, poor working conditions and violations of land rights among others.

Fixed social criteria had been proposed by the EU, however they were dropped from the final Directive as doubts remained whether they were in line with the rules of the World Trade Organisation (WTO). It is also possible that the current sustainability criteria described in the Directive are also in conflict with the rules of the WTO when applied to imports. No claim has yet been made against the EU; however it is thought that aspects of the Directive may leave the EU exposed to a challenge before the WTO.

Although there are no direct mechanisms provided to control potential social impacts in the current Directive, it does require the Commission and Member States to report and take action on these issues. The Commission is required to report biannually to the European Parliament and the Council on: the impact of social sustainability in the Community and in third countries; the impact of the Community's biofuel policy on the availability of foodstuffs at affordable prices, particularly in developing countries and on wider development issues. The Commission is then required to propose and take corrective action in the event that negative effects are reported.

An important argument against the inclusion of social criteria in the sustainability criteria is the current lack of knowledge about the pathways linking causes and effects. It is difficult and unwise to base policy decisions, which aim to avoid the negative consequences of biofuels on certain criteria, when the contributing factors are not yet fully understood or quantifiable.

Indirect Land Use Change

A significant part of the summer 2008 biofuels debate focused on indirect land use change and its exclusion from the EU sustainability criteria (European Environmental Bureau, 2009). Indirect land Use Change (ILUC) is not accounted for in the Renewables Directive and therefore the emissions resulting from ILUC are not included in the greenhouse gas life cycle analysis calculations. In addition, biodiverse lands and land with high carbon content, which are protected against direct land use change, are not protected against ILUC.

The Directive does, however, require the Commission to review the impact of indirect land use change and suggest ways to minimize it by the end of 2010. It urges the Commission to propose a methodology or factor for including these indirect effects into the GHG calculation methodology. The Parliament should decide on this proposal in 2012 at the latest. Attention must also be paid to direct and indirect land use change in both national reporting and in the EU biannual reporting. The Commission is also required to consider indirect land use change during bilateral agreements.

Once again a major problem with including such criteria is the lack of understanding of the mechanisms. ILUC is an acknowledged problem, however it is difficult to quantify and the science is not yet advanced enough to allow robust policy statements to be made. Several methods of incorporating ILUC have been tabled by various parties during stakeholder workshops; so far however, the stakeholders have failed to agree on a common suitable method.

For all of the areas in which the Commission is required to report, it must, upon finding adverse impacts, suggest and adopt corrective measures. There is no direction or specification given, however, to suggest what these might or should be.

Although the EU, through its requirement for constant monitoring and assessment, aims to deal with any unknown consequences that may arise, it has in fact created an extremely complex piece of legislation which fails to provide an obvious emergency break. The Commission's obligation to report and make necessary changes adds further ambiguity to the Directive. This ambiguity ironically leads to increasing uncertainty, undermining one of the EUs key arguments for setting mandatory targets; to provide certainty for investors.

Biofuel Policy in Brazil

Historically, mandates have driven biofuel consumption in Brazil. Ethanol blending started in 1975 with the National Alcohol Program (Programa Nacional do Alcool; Proálcool) to promote domestic ethanol production. Proálcool progressively ramped up mandated ethanol blend rates from fuel suppliers to 27% by 2015.

This was accompanied by an increasing set of biodiesel blending mandates, beginning in 2003 to 11% in 2019. Taken together, nearly 40 billion liters of biofuels were consumed in Brazil in 2018, primarily derived from food crops. Further expansion is expected over the next decade, as the newly introduced RenovaBio policy introduces carbon pricing into Brazil's fuel policy and could lead to further expansion. Figure below illustrates the growth of Brazil's ethanol and biodiesel consumption and blend rates over the past decade.

Brazil has been a world leader in ethanol production from sugarcane. The ethanol mandate has increased from a 4.5% blend in 1977 to 27% in 2015, resulting in more than 33 billion liters of ethanol produced in 2018.3 of that total, roughly 23 billion liters were hydrous ethanol, which can be consumed by flex-fuel vehicles. Table below compares Brazil's blend rates to those of other countries and jurisdictions with substantial biofuel policies.

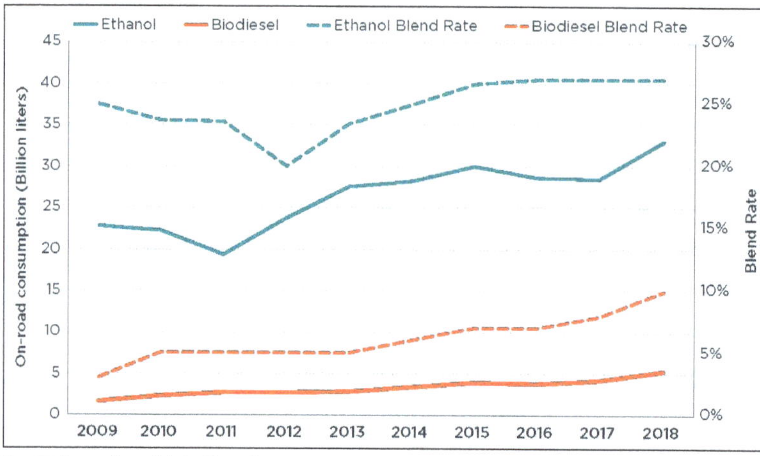

Fig: Ethanol and biodiesel consumption and blending in Brazil, 2009–2018.

Country/region	Ethanol blend rate	Biodiesel blend rate
United States	10%	5%
EU-28	5.7%	7.2%
Brazil	27%	11%
Indonesia	0%	20%

Brazil's ethanol use mandate, Proálcool, primarily supports its production through a series of federal and regional tax incentives. At the federal level, Brazil levies a tax on gasoline through the Contribution for Intervention in Economic Domain (CIDE) and Social Integration Program/Contribution for Financing Social Security (PIS/COFINS) policies; in early 2019, these policies have totaled R$ 0.8925 per liter. Although ethanol is not taxed under CIDE, ethanol producers are taxed under the PIS/COFINS policy at a rate of R$ 0.1309 per liter in March, 2019. The difference between the total tax rates for gasoline and ethanol increases the competitive advantage of ethanol, even after accounting for the difference in energy density between the two fuels. Depending on the regional tax rates and fluctuations in the price of crude oil, ethanol may have a competitive advantage relative to gasoline and vice versa.

The wide-scale deployment of flex-fuel vehicles capable of using higher ethanol blends in Brazil's vehicle fleet ensures a large market for high-ethanol blends. The Inovar-Auto fiscal incentive program has promoted increased vehicle efficiency through a vehicle energy consumption target (with penalties for noncompliance) in conjunction with credit multipliers and tax breaks for manufacturers of vehicles with advanced propulsion systems. Through the program, both ethanol and flex-fuel vehicles produced in Brazil have qualified for a tax break on the industrialized products tax and can qualify for off-cycle credits to meet the energy consumption target. The Rota 2030 program, set to replace Inovar-Auto, will continue to provide tax credits to flex-fuel and ethanol vehicles through 2032. As a result of strong policy support, flex-fuel vehicles constituted approximately 98% of all passenger cars in 2018 and approximately two-thirds of the light-duty vehicle fleet; 100% ethanol (E100)–powered vehicles are estimated to constitute 0.7% of the total vehicle fleet.

The biodiesel industry has seen a rapid expansion over the past decade. Beginning in 2003, the National Program of Biodiesel Production and Use (NPBP) has implemented a biodiesel mandate that has ramped up quickly from 2% in 2008 to 10% in 2018. Brazil's National Council for Energy Policies (CNPE) has proposed a 1% annual increase to reach B15 (15% biodiesel) blending rates by 2023 as long as certain technical requirements are met. The National Agency of Petroleum, Natural Gas and Biofuels (ANP) has already established specifications for biofuel blends up to 30% and has even established criteria for the use of B100 blends on an experimental basis.

Government support for biodiesel is ensured through a price floor established through an auction system. The federal government, via ANP, operates regular public reverse auctions where biodiesel producers competitively bid to sell lots of biodiesel to buyers. Obligated parties for the blending mandate, such as blenders of fossil diesel, are then required to pay the prices set by the auction for their biodiesel lots.

The rapid expansion of Brazil's biodiesel industry corresponds with the growth and expansion of its soy industry, which provides more than two-thirds of its biodiesel. Roughly 12.5% comes from tallow. Although biodiesel blend rates have increased in recent years, there has not been a corresponding increase in compatible vehicles. The biodiesel mandate stands at 11% in 2019, with even higher allowed blend rates for rail, agricultural, and industrial uses. Local-level initiatives could increase sales of vehicles designed for high-biodiesel blends, such as the São Paulo Climate Committee's goal of transitioning to a fossil-free bus fleet. In addition, the city of São Paulo revised its Climate Change Law in January 2018 that sets 10-year and 20-year targets for fleet-wide reductions in tailpipe emissions of fossil carbon dioxide (CO_2) and the air pollutants particulate matter (PM) and nitrogen oxides (NO_x).

The RenovaBio policy builds on Brazil's existing ethanol blending mandate to include a 10% carbon intensity (CI) reduction target for the transport fuel mix, relative to its 2017 baseline of 74.25 gCO2e per MJ of fuel by 2028. The policy would establish tradeable decarbonization (CBIO) credits generated by the production of biofuels and then redeemed by fuel blenders, increasing the cost of fossil fuels while reducing the relative costs of biofuels. Over the past decade, the low price of oil, in conjunction with climate and economic instability, has resulted in some stagnation within Brazil's domestic ethanol industry. The introduction of the RenovaBio program in 2018 was intended to both support the expansion of ethanol production and improve the cost proposition for ethanol relative to gasoline. The Ministry of Mines and Energy estimates that with the added incentives, the overall demand for Brazilian ethanol will increase to 45 billion liters by 2027.

The RenovaBio policy uses life-cycle greenhouse gas (GHG) accounting to evaluate the climate impacts of fuels but does not account for any emissions from indirect land-use change (ILUC). ILUC emissions are those attributable to market-mediated land expansion caused by biofuel demand for crop-derived feedstocks. Because these emissions may undermine the carbon savings associated with biofuel use, policies such as the U.S. Renewable Fuel Standard (RFS) and California's Low Carbon Fuel Standard (LCFS) assign crop-derived fuels an ILUC emission factor estimated through economic modeling. RenovaBio incorporates sustainability criteria to avoid the use of biofuels directly grown either on high–carbon stock land or in areas tied to deforestation. Although those land protections may discourage direct deforestation for biofuel production, they may do little to prevent indirect cropland expansion caused by greater demand for food crops.

6

Various Issues Related to Biofuels

There are numerous issues which are studied in relation to biofuels. Some of the prominent issues are the environmental impact of biodiesel, deforestation and indirect land use change impacts of biofuels. The topics elaborated in this chapter will help in gaining a better perspective about these issues connected to biofuels.

Issues Relating to Biofuels

There are various social, economic, environmental and technical issues with biofuel production and use, which have been discussed in the popular media and scientific journals. These include: the effect of moderating oil prices, the "food vs fuel" debate, poverty reduction potential, carbon emissions levels, sustainable biofuel production, deforestation and soil erosion, loss of biodiversity, effect on water resources, the possible modifications necessary to run the engine on biofuel, as well as energy balance and efficiency. The International Resource Panel, which provides independent scientific assessments and expert advice on a variety of resource-related themes, assessed the issues relating to biofuel use in its first report *Towards sustainable production and use of resources: Assessing Biofuels*. In it, it outlined the wider and interrelated factors that need to be considered when deciding on the relative merits of pursuing one biofuel over another. It concluded that not all biofuels perform equally in terms of their effect on climate, energy security and ecosystems, and suggested that environmental and social effects need to be assessed throughout the entire life-cycle.

Social and Economic Effects

Oil Price Moderation

The International Energy Agency's *World Energy Outlook 2006* concludes that rising oil demand, if left unchecked, would accentuate the consuming countries' vulnerability to a severe supply disruption and resulting price shock. The report suggested that biofuels may one day offer a viable alternative, but also that "the implications of the use of biofuels for global security as well as for economic, environmental, and public health need to be further evaluated".

According to Francisco Blanch, a commodity strategist for Merrill Lynch, crude oil would be trading 15 per cent higher and gasoline would be as much as 25 per cent more expensive, if it were not for biofuels. Gordon Quaiattini, president of the Canadian Renewable Fuels Association, argued that a healthy supply of alternative energy sources will help to combat gasoline price spikes.

"Food vs. Fuel" Debate

Food vs fuel is the debate regarding the risk of diverting farmland or crops for biofuels production in detriment of the food supply on a global scale. Essentially the debate refers to the possibility that by farmers increasing their production of these crops, often through government subsidy incentives, their time and land is shifted away from other types of non-biofuel crops driving up the price of non-biofuel crops due to the decrease in production. Therefore, it is not only that there is an increase in demand for the food staples, like corn and cassava, that sustain the majority of the world's poor but this also has the potential to increase the price of the remaining crops that these individuals would otherwise need to utilize to supplement their diets. A recent study for the International Centre for Trade and Sustainable Development shows that market-driven expansion of ethanol in the US increased maize prices by 21 percent in 2009, in comparison with what prices would have been had ethanol production been frozen at 2004 levels. A November 2011 study states that biofuels, their production, and their subsidies are leading causes of agricultural price shocks. The counter-argument includes considerations of the type of corn that is utilized in biofuels, often field corn not suitable for human consumption; the portion of the corn that is used in ethanol, the starch portion; and the negative effect higher prices for corn and grains have on government welfare for these products. The "food vs. fuel" or "food or fuel" debate is internationally controversial, with disagreement about how significant this is, what is causing it, what the effect is, and what can or should be done about it.

Poverty Reduction Potential

Researchers at the Overseas Development Institute have argued that biofuels could help to reduce poverty in the developing world, through increased employment, wider economic growth multipliers and by stabilising oil prices (many developing countries are net importers of oil). However, this potential is described as 'fragile', and is reduced where feedstock production tends to be large scale, or causes pressure on limited agricultural resources: capital investment, land, water, and the net cost of food for the poor.

With regards to the potential for poverty reduction or exacerbation, biofuels rely on many of the same policy, regulatory or investment shortcomings that impede agriculture as a route to poverty reduction. Since many of these shortcomings require policy improvements at a country level rather than a global one, they argue for a country-by-country analysis of the potential poverty effects of biofuels. This would consider, among other things, land administration systems, market coordination and prioritizing investment in biodiesel, as this 'generates more labour, has lower transportation costs and uses simpler technology'. Also necessary are reductions in the tariffs on biofuel imports regardless of the country of origin, especially due to the increased efficiency of biofuel production in countries such as Brazil.

Sustainable Biofuel Production

Responsible policies and economic instruments would help to ensure that biofuel commercialization, including the development of new cellulosic technologies, is sustainable. Responsible commercialization of biofuels represents an opportunity to enhance sustainable economic prospects in Africa, Latin America and impoverished Asia.

Environmental Effects

Soil Erosion and Deforestation

Large-scale deforestation of mature trees (which help remove CO_2 through photosynthesis — much better than sugar cane or most other biofuel feedstock crops do) contributes to soil erosion, unsustainable global warming atmospheric greenhouse gas levels, loss of habitat, and a reduction of valuable biodiversity (both on land as in oceans). Demand for biofuel has led to clearing land for palm oil plantations. In Indonesia alone, over 9,400,000 acres (38,000 km^2) of forest have been converted to plantations since 1996.

A portion of the biomass should be retained onsite to support the soil resource. Normally this will be in the form of raw biomass, but processed biomass is also an option. If the exported biomass is used to produce syngas, the process can be used to co-produce biochar, a low-temperature charcoal used as a soil amendment to increase soil organic matter to a degree not practical with less recalcitrant forms of organic carbon. For co-production of biochar to be widely adopted, the soil amendment and carbon sequestration value of co-produced charcoal must exceed its net value as a source of energy.

Some commentators claim that removal of additional cellulosic biomass for biofuel production will further deplete soils.

Effect on Water Resources

Increased use of biofuels puts increasing pressure on water resources in at least two ways: water use for the irrigation of crops used as feedstocks for biodiesel production; and water use in the production of biofuels in refineries, mostly for boiling and cooling.

In many parts of the world supplemental or full irrigation is needed to grow feedstocks. For example, if in the production of corn (maize) half the water needs of crops are met through irrigation and the other half through rainfall, about 860 liters of water are needed to produce one liter of ethanol. However, in the United States only 5-15% of the water required for corn comes from irrigation while the other 85-95% comes from natural rainfall.

In the United States, the number of ethanol factories has almost tripled from 50 in 2000 to about 140 in 2008. A further 60 or so are under construction, and many more are planned. Projects are being challenged by residents at courts in Missouri (where water is drawn from the Ozark Aquifer), Iowa, Nebraska, Kansas (all of which draw water from the non-renewable Ogallala Aquifer), central Illinois (where water is drawn from the Mahomet Aquifer) and Minnesota.

For example, the four ethanol crops: corn, sugarcane, sweet sorghum and pine yield net energy. However, increasing production in order to meet the U.S. Energy Independence and Security Act mandates for renewable fuels by 2022 would take a heavy toll in the states of Florida and Georgia. The sweet sorghum, which performed the best of the four, would increase the amount of freshwater withdrawals from the two states by almost 25%.

Loss of Biodiversity

Critics argue that expansion of farming for biofuel production causes unacceptable loss of biodiversity for a much less significant decrease in fossil fuel consumption. The loss of biodiversity also makes heavy dependence on biofuels very risky by reducing our ability to deal with blights affecting the few important biofuel crops. Food crops have recovered from blights when the old stock was mixed with blight resistant wild strains, but as biodiversity is lost due to excessive agriculture, the possibilities for recovering from blights are lost.

Pollution

Formaldehyde, acetaldehyde and other aldehydes are produced when alcohols are oxidized. When only a 10% mixture of ethanol is added to gasoline (as is common in American E10 gasohol and elsewhere), aldehyde emissions increase 40%. Some study results are conflicting on this fact however, and lowering the sulfur content of biofuel mixes lowers the acetaldehyde levels. Burning biodiesel also emits aldehydes and other potentially hazardous aromatic compounds which are not regulated in emissions laws.

Many aldehydes are toxic to living cells. Formaldehyde irreversibly cross-links protein amino acids, which produces the hard flesh of embalmed bodies. At high concentrations in an enclosed space, formaldehyde can be a significant respiratory irritant causing nose bleeds, respiratory distress, lung disease, and persistent headaches. Acetaldehyde, which is produced in the body by alcohol drinkers and found in the mouths of smokers and those with poor oral hygiene, is carcinogenic and mutagenic.

The European Union has banned products that contain Formaldehyde, due to its documented carcinogenic characteristics. The U.S. Environmental Protection Agency has labeled Formaldehyde as a probable cause of cancer in humans.

Brazil burns significant amounts of ethanol biofuel. Gas chromatograph studies were performed of ambient air in São Paulo, Brazil, and compared to Osaka, Japan, which does not burn ethanol fuel. Atmospheric Formaldehyde was 160% higher in Brazil, and Acetaldehyde was 260% higher.

Technical Issues

Energy Efficiency and Energy Balance

Production of biofuels from raw materials requires energy (for farming, transport and conversion to final product, and the production / application of fertilizers, pesticides, herbicides, and fungicides), and has environmental consequences.

The energy balance of a biofuel (sometimes called "Net energy gain" and EROEI) is determined by the amount of energy put into the manufacture of fuel compared to the amount of energy released when it is burned in a vehicle. This varies by feedstock and according to the assumptions used. Biodiesel made from sunflowers may produce only 0.46 times the input rate of fuel energy. Biodiesel made from soybeans may produce 3.2 times the input rate of fossil fuels. This compares to 0.805 for gasoline and 0.843 for diesel made from petroleum. Biofuels may require higher energy input per unit of BTU energy content produced than fossil fuels: petroleum can be pumped out of the ground and processed more efficiently than biofuels can be grown and processed. However, this is not necessarily a reason to use oil instead of biofuels, nor does it affect the environmental benefits provided by a given biofuel.

Studies have been done that calculate energy balances for biofuel production. Some of these show large differences depending on the biomass feedstock used and location.

To explain one specific example, a June 17, 2006 editorial in the Wall. St. Journal stated, "The most widely cited research on this subject comes from Cornell's David Pimental and Berkeley's Ted Patzek. They've found that it takes more than a gallon of fossil fuel to make one gallon of ethanol — 29% more. That's because it takes enormous amounts of fossil-fuel energy to grow corn (using fertilizer and irrigation), to transport the crops and then to turn that corn into ethanol."

Life cycle assessments of biofuel production show that under certain circumstances, biofuels produce only limited savings in energy and greenhouse gas emissions. Fertilizer inputs and transportation of biomass across large distances can reduce the greenhouse gas (GHG) savings achieved. The location of biofuel processing plants can be planned to minimize the need for transport, and agricultural regimes can be developed to limit the amount of fertiliser used for biomass production. A European study on the greenhouse gas emissions found that well-to-wheel (WTW) CO_2 emissions of biodiesel from seed crops such as rapeseed could be almost as high as fossil diesel. It showed a similar result for bio-ethanol from starch crops, which could have almost as many WTW CO_2 emissions as fossil petrol. This study showed that second generation biofuels have far lower WTW CO_2 emissions.

Other independent LCA studies show that biofuels save around 50% of the CO_2 emissions of the equivalent fossil fuels. This can be increased to 80-90% GHG emissions savings if second generation processes or reduced fertiliser growing regimes are used. Further GHG savings can be achieved by using by-products to provide heat, such as using bagasse to power ethanol production from sugarcane.

Collocation of synergistic processing plants can enhance efficiency. One example is to use the exhaust heat from an industrial process for ethanol production, which can then recycle cooler processing water, instead of evaporating hot water that warms the atmosphere.

Biomass planting mandated by law (as in European Union) results in large quantities of biomass being transported to EU from Africa, Asia and Americas (Canada, USA, Brazil). For example, in Poland as much as 85% of biomass used is imported from outside of EU, with single electric plant in Łódź importing over 7'000 tons of wood biomass from Republic of Komi (Russia) over distance of 7'000 kilometers on monthly basis.

T. A. Kiefer of the US Air Force Air War College, in a paper entitled "The False Promise of Liquid Biofuels", laid out factors that preclude biofuels from replacing petroleum as a national-scale transportation fuel. Kiefer states "The energy content of the final-product biofuel compared to the energy required to produce it proves to be a very poor investment, especially compared to other alternatives. In many cases, there is net loss of energy." He concludes "...pursuit of biofuels creates irreversible harm to the environment, increases greenhouse gas emissions, undermines food security, and promotes abuse of human rights."

Solar Energy Efficiency

Biofuels from plant materials convert energy that was originally captured from solar energy via photosynthesis. A comparison of conversion efficiency from solar to usable energy (taking into account the whole energy budgets) shows that photovoltaics are 100 times more efficient than corn ethanol and 10 times more efficient than the best biofuel. However, photovoltaics produce electricity rather than storable, portable liquid hydrocarbon fuel, so they are largely irrelevant for powering the large existing fleet of vehicles and equipment having internal combustion engines. Also from the economic point of view, green plants are self-assembling organisms and therefore much cheaper to produce than photovoltaic cells.

Carbon Emissions

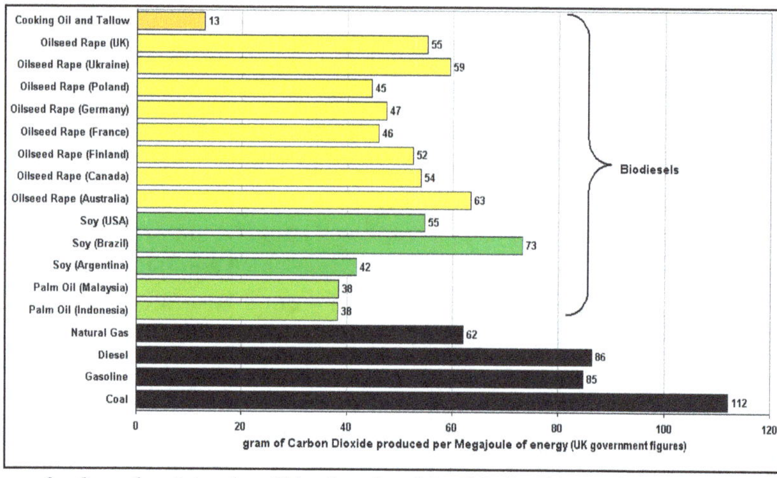

Graph of UK figures for the carbon intensity of bioethanol and fossil fuels. This graph assumes that all bioethanols are burnt in their country of origin and that previously existing cropland is used to grow the feedstock.

Biofuels and other forms of renewable energy aim to be carbon neutral or even carbon negative. Carbon neutral means that the carbon released during the use of the fuel, e.g. through burning to power transport or generate electricity, is reabsorbed and balanced by the carbon absorbed by new plant growth. These plants are then harvested to make the next batch of fuel. Carbon neutral fuels lead to no net increases in human contributions to atmospheric carbon dioxide levels, reducing the human contributions to global warming. A carbon negative aim is achieved when a portion of the biomass is used for carbon sequestration. Calculating exactly how much greenhouse gas (GHG) is produced in burning biofuels is a complex and inexact process, which depends very much on the method by which the fuel is produced and other assumptions made in the calculation.

The carbon emissions (carbon footprint) produced by biofuels are calculated using a technique called Life Cycle Analysis (LCA). This uses a "cradle to grave" or "well to wheels" approach to calculate the total amount of carbon dioxide and other greenhouse gases emitted during biofuel production, from putting seed in the ground to using the fuel in cars and trucks. Many different LCAs have been done for different biofuels, with widely differing results. Several well-to-wheel analysis for biofuels has shown that first generation biofuels can reduce carbon emissions, with savings depending on the feedstock used, and second generation biofuels can produce even higher savings when compared to using fossil fuels. However, those studies did not take into account emissions from nitrogen fixation, or additional carbon emissions due to indirect land use changes. In addition, many LCA studies fail to analyze the effect of substitutes that may come into the market to replace current biomass-based products. In the case of Crude Tall Oil, a raw material used in the production of pine chemicals and now being diverted for use in biofuel, an LCA study found that the global carbon footprint of pine chemicals produced from CTO is 50 percent lower than substitute products used in the same situation offsetting any gains from utilizing a biofuel to replace fossil fuels. Additionally the study showed that fossil fuels are not reduced when CTO is diverted to biofuel use and the substitute products consume disproportionately more energy. This diversion will negatively affect an industry that contributes significantly to the world economy, globally producing more than 3 billion pounds of pine chemicals annually in complex, high technology refineries and providing jobs directly and indirectly for tens of thousands of workers.

A paper published in February 2008 in Sciencexpress by a team led by Searchinger from Princeton University concluded that once considered indirect land use changes effects in the life cycle assessment of biofuels used to substitute gasoline, instead of savings both corn and cellulosic ethanol increased carbon emissions as compared to gasoline by 93 and 50 percent respectively. A second paper published in the same issue of Sciencexpress, by a team led by Fargione from The Nature Conservancy, found that a carbon debt is created when natural lands are cleared and being converted to biofuel production and to crop production when agricultural land is diverted to biofuel production, therefore this carbon debt applies to both direct and indirect land use changes.

The Searchinger and Fargione studies gained prominent attention in both the popular media and in scientific journals. The methodology, however, drew some criticism, with Wang and Haq from Argonne National Laboratory posted a public letter and send their criticism about the Searchinger paper to Letters to Science. Another criticism by Kline and Dale from Oak Ridge National Laboratory was published in Letters to Science. They argued that Searchinger et al. and Fargione et al. "*...do not provide adequate support for their claim that biofuels cause high emissions due to land-use change*. The U.S. biofuel industry also reacted, claiming in a public letter, that the "*Searchinger study is clearly a "worst case scenario" analysis...*" and that this study "*relies on a long series of highly subjective assumptions...*".

Modifications Necessary to Internal Combustion Engines

The modifications necessary to run internal combustion engines on biofuel depend on the type of biofuel used, as well as the type of engine used. For example, gasoline engines can run without any modification at all on biobutanol. Minor modifications are however needed to run on bioethanol or biomethanol. Diesel engines can run on the latter fuels, as well as on vegetable oils (which are cheaper). However, the latter is only possible when the engine has been foreseen with indirect injection. If no indirect injection is present, the engine hence needs to be fitted with this.

Campaigns

A number of environmental NGOs campaign against the production of biofuels as a large-scale alternative to fossil fuels. For example, Friends of the Earth state that "the current rush to develop agrofuels (or biofuels) on a large scale is ill-conceived and will contribute to an already unsustainable trade whilst not solving the problems of climate change or energy security". Some mainstream environmental groups support biofuels as a significant step toward slowing or stopping global climate change. However, supportive environmental groups generally hold the view that biofuel production can threaten the environment if it is not done sustainably. This finding has been backed by reports of the UN, the IPCC, and some other smaller environmental and social groups as the EEB and the Bank Sarasin, which generally remain negative about biofuels.

As a result, governmental and environmental organizations are turning against biofuels made in a non-sustainable way (hereby preferring certain oil sources as jatropha and lignocellulose over palm oil) and are asking for global support for this. Also, besides supporting these more sustainable biofuels, environmental organizations are redirecting to new technologies that do not use internal combustion engines such as hydrogen and compressed air.

Several standard-setting and certification initiatives have been set up on the topic of biofuels. The "Roundtable on Sustainable Biofuels" is an international initiative which brings together farmers, companies, governments, non-governmental organizations, and scientists who are interested in the sustainability of biofuels production and distribution. During 2008, the Roundtable is developing a series of principles and criteria for sustainable biofuels production through meetings, teleconferences, and online discussions. In a similar vein, the Bonsucro standard has been developed as a metric-based certificate for products and supply chains, as a result of an ongoing multi-stakeholder initiative focussing on the products of sugar cane, including ethanol fuel.

The increased manufacture of biofuels will require increasing land areas to be used for agriculture. Second and third generation biofuel processes can ease the pressure on land, because they can use waste biomass, and existing (untapped) sources of biomass such as crop residues and potentially even marine algae.

In some regions of the world, a combination of increasing demand for food, and increasing demand for biofuel, is causing deforestation and threats to biodiversity. The best reported example of this is the expansion of oil palm plantations in Malaysia and Indonesia, where rainforest is being destroyed to establish new oil palm plantations. It is an important fact that 90% of the palm oil produced in Malaysia is used by the food industry; therefore biofuels cannot be held solely responsible for this deforestation. There is a pressing need for sustainable palm oil production for the food and fuel industries; palm oil is used in a wide variety of food products. The *Roundtable on Sustainable Biofuels* is working to define criteria, standards and processes to promote sustainably produced biofuels. Palm oil is also used in the manufacture of detergents, and in electricity and heat generation both in Asia and around the world (the UK burns palm oil in coal-fired power stations to generate electricity).

Significant area is likely to be dedicated to sugar cane in future years as demand for ethanol increases worldwide. The expansion of sugar cane plantations will place pressure on environmentally sensitive native ecosystems including rainforest in South America. In forest ecosystems, these effects themselves will undermine the climate benefits of alternative fuels, in addition to representing a major threat to global biodiversity.

Although biofuels are generally considered to improve net carbon output, biodiesel and other fuels do produce local air pollution, including nitrogen oxides, the principal cause of smog.

Politics

Steven Rattner, former "auto czar" for U.S. President Barack Obama, wrote an Op-ed for *The New York Times* in June, 2011, entitled "The Great Corn Con," characterizing ethanol as "an example of government policy run amok." Along with the economic and environmental effects of the U.S. policy, he noted the effect of the issue on presidential politics;

Those [presidential] hopefuls have seen no need for a foolish consistency. John McCain and John Kerry were against ethanol subsidies, then as candidates were for them. Having lost the presidency, Mr. McCain is now against them again. Al Gore was for ethanol before he was against it. This time, one hopeful is experimenting with counter-programming: as governor of corn-producing Minnesota, Tim Pawlenty pushed for subsidies before he embraced a "straight talk" strategy.

Rattner did not address President Obama's long-time alignment with Illinois and U.S. corn producers on the issue.

In April, 2014 an article in the Financial Times described how biofuel manufacturers in Europe felt threatened by changing European Union climate policies. After mapping rigorous regulations for cleaner fuels in 2009, the European Commission decided not to set specific fuel targets from 2020 to 2030. Instead the Commission recommended a "more holistic and integrated approach" to creating an efficient biofuels policy. The EC also called for "an improved biomass policy" to "maximize the resource efficient use of biomass in order to deliver robust and verifiable greenhouse gas savings and to allow for fair competition between the various uses of biomass resources in the construction sector, paper and pulp industries and biochemical and energy production." Fair competition in the acquisition of biomass feedstocks is what established bio-industries are asking for. Incentives provided to biofuel producers create an uneven playing field. "The pine chemical industry, which uses a co product called crude tall oil from the paper-pulping process, says it's concerned that incentives will divert its major raw material into biofuel production," stated a Greenwire article.

Food vs. Fuel

An ethanol fuel plant under construction, Butler County, Iowa.

Food versus fuel is the dilemma regarding the risk of diverting farmland or crops for biofuels production to the detriment of the food supply. The biofuel and food price debate involves wide-ranging views, and is a long-standing, controversial one in the literature. There is disagreement about the significance of the issue, what is causing it, and what can or should be done to remedy the situation. This complexity and uncertainty is due to the large number of impacts and feedback loops that can positively or negatively affect the price system. Moreover, the relative strengths of these positive and negative impacts vary in the short and long terms, and involve delayed effects. The academic side of the debate is also blurred by the use of different economic models and competing forms of statistical analysis.

Biofuel production has increased in recent years. Some commodities like maize (corn), sugar cane or vegetable oil can be used either as food, feed, or to make biofuels. For example, since 2006, a portion of land that was also formerly used to grow other crops in the United States is now used to grow corn for biofuels, and a larger share of corn is destined to ethanol production, reaching 25% in 2007. Second generation biofuels could potentially combine farming for food and fuel and moreover, electricity could be generated simultaneously, which could be beneficial for developing countries and rural areas in developed countries. With global demand for biofuels on the increase due to the oil price increases taking place since 2003 and the desire to reduce oil dependency as well as reduce GHG emissions from transportation, there is also fear of the potential destruction of natural habitats by being converted into farmland. Environmental groups have raised concerns about this trade-off for several years, but now the debate reached a global scale due to the 2007–2008 world food price crisis. On the other hand, several studies do show that biofuel production can be significantly increased without increased acreage. Therefore, stating that the crisis in hand relies on the food scarcity.

Biofuels are not a new phenomenon. Before the industrialisation, horses were the primary (and humans probably the secondary) source of power for transportation and physical work, requiring food. The growing of crops for horses (typically oat) for carrying out physical work is of course comparable to the growing of crops for biofuels for engines, albeit on a smaller scale, because production since then has increased.

Brazil has been considered to have the world's first sustainable biofuels economy and its government claims Brazil's sugar cane based ethanol industry has not contributed to the 2008 food crisis. A World Bank policy research working paper released in July 2008 concluded that "...large increases in biofuels production in the United States and Europe are the main reason behind the steep rise in global food prices", and also stated that "Brazil's sugar-based ethanol did not push food prices appreciably higher". However, a 2010 study also by the World Bank concluded that their previous study may have overestimated the contribution of biofuel production, as "the effect of biofuels on food prices has not been as large as originally thought, but that the use of commodities by financial investors (the so-called "financialisation of commodities") may have been partly responsible for the 2007/08 spike." A 2008 independent study by OECD also found that the impact of biofuels on food prices are much smaller.

Food Price Inflation

From 1974 to 2005 real food prices (adjusted for inflation) dropped by 75%. Food commodity prices were relatively stable after reaching lows in 2000 and 2001. Therefore, recent rapid food price increases are considered extraordinary. A World Bank policy research working paper published on July 2008 found that the increase in food commodities prices was led by grains, with sharp price increases in 2005 despite record crops worldwide. From January 2005 until June 2008, maize prices almost tripled, wheat increased 127 percent, and rice rose 170 percent. The increase in grain prices was followed by increases in fats and oil prices in mid-2006. On the other hand, the study found that sugar cane production has increased rapidly, and it was large enough to keep sugar price increases small except for 2005 and early 2006. The paper concluded that biofuels produced from grains have raised food prices in combination with other related factors between 70 to 75 percent, but ethanol produced from sugar cane has not contributed significantly to the recent increase in food commodities prices.

An economic assessment report published by the OECD in July 2008 found that "...the impact of current biofuel policies on world crop prices, largely through increased demand for cereals and vegetable oils, is significant but should not be overestimated. Current biofuel support measures alone are estimated to increase average wheat prices by about 5 percent, maize by around 7 percent and vegetable oil by about 19 percent over the next 10 years."

Corn is used to make ethanol and prices went up by a factor of three in less than 3 years (measured in US dollars). Reports in 2007 linked stories as diverse as food riots in Mexico due to rising prices of corn for tortillas, and reduced profits at Heineken the large international brewer, to the increasing use of corn (maize) grown in the US Midwest for ethanol production. (In the case of beer, the barley area was cut in order to increase corn production. Barley is not currently used to produce ethanol.) Wheat is up by almost a factor of 3 in 3 years, while soybeans are up by a factor of 2 in 2 years (both measured in US dollars).

As corn is commonly used as feed for livestock, higher corn prices lead to higher prices in animal source foods. Vegetable oil is used to make biodiesel and has about doubled in price in the last couple years. The price is roughly tracking crude oil prices. The 2007–2008 world food price crisis is blamed partly on the increased demand for biofuels. During the same period rice prices went up by a factor of 3 even though rice is not directly used in biofuels.

The USDA expects the 2008/2009 wheat season to be a record crop and 8% higher than the previous year. They also expect rice to have a record crop. Wheat prices have dropped from a high over $12/bushel in May 2008 to under $8/bushel in May. Rice has also dropped from its highs.

According to a 2008 report from the World Bank the production of biofuel pushed food prices up. These conclusions were supported by the Union of Concerned Scientists in their September 2008 newsletter in which they remarked that the World Bank analysis "contradicts U.S. Secretary of Agriculture Ed Schaffer's assertion that biofuels account for only a small percentage of rising food prices."

According to the October Consumer Price Index released Nov. 19, 2008, food prices continued to rise in October 2008 and were 6.3 percent higher than October 2007. Since July 2008 fuel costs dropped by nearly 60 percent.

Proposed Causes

Ethanol Fuel as an Oxygenate Additive

The demand for ethanol fuel produced from field corn was spurred in the U.S. by the discovery that methyl tertiary butyl ether (MTBE) was contaminating groundwater. MTBE use as an oxygenate additive was widespread due to mandates of the Clean Air Act amendments of 1992 to reduce carbon monoxide emissions. As a result, by 2006 MTBE use in gasoline was banned in almost 20 states. There was also concern that widespread and costly litigation might be taken against the U.S. gasoline suppliers, and a 2005 decision refusing legal protection for MTBE, opened a new market for ethanol fuel, the primary substitute for MTBE. At a time when corn prices were around US$2 a bushel, corn growers recognized the potential of this new market and delivered accordingly. This demand shift took place at a time when oil prices were already significantly rising.

Other Factors

That food prices went up at the same time fuel prices went up is not surprising and should not be entirely blamed on biofuels. Energy costs are a significant cost for fertilizer, farming, and food distribution. Also, China and other countries have had significant increases in their imports as their economies have grown. Sugar is one of the main feedstocks for ethanol and prices are down from 2 years ago. Part of the food price increase for international food commodities measured in US dollars is due to the dollar being devalued. Protectionism is also an important contributor to price increases. 36% of world grain goes as fodder to feed animals, rather than people.

Over long time periods population growth and climate change could cause food prices to go up. However, these factors have been around for many years and food prices have jumped up in the last 3 years, so their contribution to the current problem is minimal.

Government Regulations of Food and Fuel Markets

France, Germany, the United Kingdom and the United States governments have supported biofuels with tax breaks, mandated use, and subsidies. These policies have the unintended consequence of diverting resources from food production and leading to surging food prices and the potential destruction of natural habitats.

Fuel for agricultural use often does not have fuel taxes (farmers get duty-free petrol or diesel fuel). Biofuels may have subsidies and low/no retail fuel taxes. Biofuels compete with retail gasoline and diesel prices which have substantial taxes included. The net result is that it is possible for a farmer to use more than a gallon of fuel to make a gallon of biofuel and still make a profit. Some argue that this is a bad distortion of the market. There have been thousands of scholarly papers analyzing how much energy goes into making ethanol from corn and how that compares to the energy in the ethanol.

A World Bank policy research working paper concluded that food prices have risen by 35 to 40 percent between 2002 and 2008, of which 70 to 75 percent is attributable to biofuels. The "month-by-month" five-year analysis disputes that increases in global grain consumption and droughts were responsible for significant price increases, reporting that this had only a marginal impact. Instead the report argues that the EU and US drive for biofuels has had by far the biggest impact on food supply and prices, as increased production of biofuels in the US and EU were supported by subsidies and tariffs on imports, and considers that without these policies, price increases would have been smaller. This research also concluded that Brazil's sugar cane based ethanol has not raised sugar prices significantly, and recommends removing tariffs on ethanol imports by both the US and EU, to allow more efficient producers such as Brazil and other developing countries, including many African countries, to produce ethanol profitably for export to meet the mandates in the EU and the US.

An economic assessment published by the OECD in July 2008 agrees with the World Bank report recommendations regarding the negative effects of subsidies and import tariffs, but found that the estimated impact of biofuels on food prices are much smaller. The OECD study found that trade restrictions, mainly through import tariffs, protect the domestic industry from foreign competitors but impose a cost burden on domestic biofuel users and limits alternative suppliers. The report is also critical of limited reduction of GHG emissions achieved from biofuels based on feedstocks used in Europe and North America, finding that the current biofuel support policies would reduce greenhouse gas emissions from transport fuel by no more than 0.8% by 2015, while Brazilian ethanol from sugar cane reduces greenhouse gas emissions by at least 80% compared to fossil fuels. The assessment calls for the need for more open markets in biofuels and feedstocks in order to improve efficiency and lower costs.

Oil Price Increases

Oil price increases since 2003 resulted in increased demand for biofuels. Transforming vegetable oil into biodiesel is not very hard or costly so there is a profitable arbitrage situation if vegetable oil is much cheaper than diesel. Diesel is also made from crude oil, so vegetable oil prices are partially linked to crude oil prices. Farmers can switch to growing vegetable oil crops if those are more profitable than food crops. So all food prices are linked to vegetable oil prices, and in turn to crude oil prices. A World Bank study concluded that oil prices and a weak dollar explain 25-30% of total price rise between January 2002 until June 2008.

Demand for oil is outstripping the supply of oil and oil depletion is expected to cause crude oil prices to go up over the next 50 years. Record oil prices are inflating food prices worldwide, including those crops that have no relation to biofuels, such as rice and fish.

In Germany and Canada it is now much cheaper to heat a house by burning grain than by using fuel derived from crude oil. With oil at $120/barrel a savings of a factor of 3 on heating costs is possible. When crude oil was at $25/barrel there was no economic incentive to switch to a grain fed heater.

From 1971 to 1973, around the time of the 1973 oil crisis, corn and wheat prices went up by a factor of 3. There was no significant biofuel usage at that time.

US Government Policy

Some argue that the US government policy of encouraging ethanol from corn is the main cause for food price increases. US Federal government ethanol subsidies total $7 billion per year, or $1.90 per gallon. Ethanol provides only 55% as much energy as gasoline per gallon, realizing about a $3.45 per gallon gasoline trade off. Corn is used to feed chickens, cows, and pigs, so higher corn prices lead to higher prices for chicken, beef, pork, milk, cheese, etc.

U.S. Senators introduced the *BioFuels Security Act* in 2006. "It's time for Congress to realize what farmers in America's heartland have known all along - that we have the capacity and ingenuity to decrease our dependence on foreign oil by growing our own fuel," said U.S. Senator for Illinois Barack Obama.

Two-thirds of U.S. oil consumption is due to the transportation sector. The Energy Independence and Security Act of 2007 has a significant impact on U.S. Energy Policy. With the high profitability of growing corn, more and more farmers switch to growing corn until the profitability of other crops goes up to match that of corn. So the ethanol/corn subsidies drive up the prices of other farm crops.

The US - an important export country for food stocks - will convert 18% of its grain output to ethanol in 2008. Across the US, 25% of the whole corn crop went to ethanol in 2007. The percentage of corn going to biofuel is expected to go up.

Since 2004 a US subsidy has been paid to companies that blend biofuel and regular fuel. The European biofuel subsidy is paid at the point of sale. Companies import biofuel to the US, blend 1% or even 0.1% regular fuel, and then ship the blended fuel to Europe, where it can get a second subsidy. These blends are called B99 or B99.9 fuel. The practice is called "splash and dash". The imported fuel may even come from Europe to the US, get 0.1% regular fuel, and then go back to Europe. For B99.9 fuel the US blender gets a subsidy of $0.999 per gallon. The European biodiesel producers have urged the EU to impose punitive duties on these subsidized imports. In 2007, US lawmakers were also looking at closing this loophole.

Freeze on First Generation Biofuel Production

The prospects for the use of biofuels could change in a relatively dramatic way in 2014. Petroleum trade groups petitioned the EPA in August 2013 to take into consideration a reduction of renewable biofuel content in transportation fuels. On November 15, 2013 the United States EPA announced a review of the proportion of ethanol that should be required by regulation. The standards established by the Energy Independence and Security Act of 2007 could be modified significantly. The announcement allows sixty days for the submission of commentary about the proposal.

Journalist George Monbiot has argued for a 5-year freeze on biofuels while their impact on poor communities and the environment is assessed. It has been suggested that a problem with Monbiot's approach is that economic drivers may be required in order to push through the development of more sustainable second-generation biofuel processes: it is possible that these could be stalled if biofuel production decreases. Some environmentalists are suspicious that second-generation biofuels may not solve the problem of a potential clash with food as they also use significant agricultural resources such as water.

A recent UN report on biofuel also raises issues regarding food security and biofuel production. Jean Ziegler, then UN Special Rapporteur on food, concluded that while the argument for biofuels in terms of energy efficiency and climate change are legitimate, the effects for the world's hungry of transforming wheat and maize crops into biofuel are "absolutely catastrophic," and terms such use of arable land a "crime against humanity." Ziegler also calls for a 5-year moratorium on biofuel production. Ziegler's proposal for a five-year ban was rejected by the U.N. Secretary Ban Ki-moon, who called for a comprehensive review of the policies on biofuels, and said that "just criticising biofuel may not be a good solution".

Food surpluses exist in many developed countries. For example, the UK wheat surplus was around 2 million tonnes in 2005. This surplus alone could produce sufficient bioethanol to replace around 2.5% of the UK's petroleum consumption, without requiring any increase in wheat cultivation or reduction in food supply or exports. However, above a few percent, there would be direct competition between first generation biofuel production and food production. This is one reason why many view second generation biofuels as increasingly important.

Non-food Crops for Biofuel

There are different types of biofuels and different feedstocks for them, and it has been proposed that only non-food crops be used for biofuel. This avoids direct competition for commodities like corn and edible vegetable oil. However, as long as farmers are able to derive a greater profit by switching to biofuels, they will. The law of supply and demand predicts that if fewer farmers are producing food the price of food will rise.

Second generation biofuels use lignocellulosic raw material such as forest residues (sometimes referred to as brown waste and black liquor from Kraft process or sulfite process pulp mills). Third generation biofuels (biofuel from algae) use non-edible raw materials sources that can be used for biodiesel and bioethanol.

Biodiesel

Soybean oil, which only represents half of the domestic raw materials available for biodiesel production in the United States, is one of many raw materials that can be used to produce biodiesel.

Non-food crops like Camelina, Jatropha, seashore mallow and mustard, used for biodiesel, can thrive on marginal agricultural land where many trees and crops won't grow, or would produce only slow growth yields. Camelina is virtually 100 percent efficient. It can be harvested and crushed for oil and the remaining parts can be used to produce high quality omega-3 rich animal feed, fiberboard, and glycerin. Camelina does not take away from land currently being utilized for food production. Most camelina acres are grown in areas that were previously not utilized for farming. For example, areas that receive limited rainfall that can not sustain corn or soybeans without the addition of irrigation can grow camelina and add to their profitability.

Jatropha cultivation provides benefits for local communities:

Cultivation and fruit picking by hand is labour-intensive and needs around one person per hectare. In parts of rural India and Africa this provides much-needed jobs - about 200,000 people worldwide now find employment through jatropha. Moreover, villagers often find that they can grow other crops in the shade of the trees. Their communities will avoid importing expensive diesel and there will be some for export too.

NBB's Feedstock Development program is addressing production of arid variety crops, algae, waste greases, and other feedstocks on the horizon to expand available material for biodiesel in a sustainable manner.

Bioalcohols

Cellulosic ethanol is a type of biofuel produced from lignocellulose, a material that comprises much of the mass of plants. Corn stover, switchgrass, miscanthus and woodchip are some of the more popular non-edible cellulosic materials for ethanol production. Commercial investment in such second-generation biofuels began in 2006/2007, and much of this investment went beyond pilot-scale plants. Cellulosic ethanol commercialization is moving forward rapidly. The world's first commercial wood-to-ethanol plant began operation in Japan in 2007, with a capacity of 1.4 million liters/year. The first wood-to-ethanol plant in the United States is planned for 2008 with an initial output of 75 million liters/year.

Other second generation biofuels may be commercialized in the future and compete less with food. Synthetic fuel can be made from coal or biomass and may be commercialized soon.

Biofuel from Food Byproducts and Coproducts

Biofuels can also be produced from the waste byproducts of food-based agriculture (such as citrus peels or used vegetable oil) to manufacture an environmentally sustainable fuel supply, and reduce waste disposal cost.

A growing percentage of U.S. biodiesel production is made from waste vegetable oil (recycled restaurant oils) and greases.

Collocation of a waste generator with a waste-to-ethanol plant can reduce the waste producer's operating cost, while creating a more-profitable ethanol production business. This innovative collocation concept is sometimes called holistic systems engineering. Collocation disposal elimination may be one of the few cost-effective, environmentally sound, biofuel strategies, but its scalability is limited by availability of appropriate waste generation sources. For example, millions of tons of wet Florida-and-California citrus peels cannot supply billions of gallons of biofuels. Due to the higher cost of transporting ethanol, it is a local partial solution, at best.

More firms are investigating the potential of fractionating technology to remove corn germ (i.e. the portion of the corn kernel that contains oil) prior to the ethanol process. Furthermore, some ethanol plants have already announced their intention to employ technology to remove the remaining vegetable oil from dried distillers grains, a coproduct of the ethanol process. Both of these technologies would add to the biodiesel raw material supply.

Biofuel Subsidies and Tariffs

Some people have claimed that ending subsidies and tariffs would enable sustainable development of a global biofuels market. Taxing biofuel imports while letting petroleum in duty-free does not fit with the goal of encouraging biofuels. Ending mandates, subsidies, and tariffs would end the distortions that current policy is causing. Some US senators advocate reducing subsidies for corn based ethanol. The US ethanol tariff and some US ethanol subsidies are currently set to expire over the next couple years. The EU is rethinking their biofuels directive due to environmental and social concerns. On January 18, 2008 the UK House of Commons Environmental Audit Committee raised similar concerns, and called for a moratorium on biofuel targets. Germany ended their subsidy of biodiesel on Jan 1 2008 and started taxing it.

Reduce Farmland Reserves and Set Asides

To avoid overproduction and to prop up farmgate prices for agricultural commodities, the EU has for a long time have had farm subsidy programs to encourage farmers not to produce and leave productive acres fallow. The 2008 crisis prompted proposals to bring some of the reserve farmland back into use, and the used area increased actually with 0.5% but today these areas are once again out of use. According to Eurostat, 18 million hectares has been abandoned since 1990, 7,4 millions hectares are currently set aside Landuse in EU and the EU has recently decided to set aside another 5-7 %, in so called Ecological Focus Areas, corresponding to 10-12 million hectares. In spite of this reduction of used land, the EU is a net exporter of e.g. wheat.

The American Bakers Association has proposed reducing the amount of farmland held in the US Conservation Reserve Program. Currently the US has 34,500,000 acres (140,000 km^2) in the program.

In Europe about 8% of the farmland is in set aside programs. Farmers have proposed freeing up all of this for farming. Two-thirds of the farmers who were on these programs in the UK are not renewing when their term expires.

Sustainable Production of Biofuels

Second generation biofuels are now being produced from the cellulose in dedicated energy crops (such as perennial grasses), forestry materials, the co-products from food production, and domestic vegetable waste. Advances in the conversion processes will almost certainly improve the sustainability of biofuels, through better efficiencies and reduced environmental impact of producing biofuels, from both existing food crops and from cellulosic sources.

Lord Ron Oxburgh suggests that responsible production of biofuels has several advantages:

Produced responsibly they are a sustainable energy source that need not divert any land from growing food nor damage the environment; they can also help solve the problems of the waste generated by Western society; and they can create jobs for the poor where previously were none. Produced irresponsibly, they at best offer no climate benefit and, at worst, have detrimental social and environmental consequences. In other words, biofuels are pretty much like any other product.

Far from creating food shortages, responsible production and distribution of biofuels represents the best opportunity for sustainable economic prospects in Africa, Latin America and impoverished Asia. Biofuels offer the prospect of real market competition and oil price moderation. Crude oil would be trading 15 per cent higher and gasoline would be as much as 25 per cent more expensive, if it were not for biofuels. A healthy supply of alternative energy sources will help to combat gasoline price spikes.

Continuation of the Status Quo

An additional policy option is to continue the current trends of government incentive for these types of crops to further evaluate the effects on food prices over a longer period of time due to the relatively recent onset of the biofuel production industry. Additionally, by virtue of the newness of the industry we can assume that like other startup industries techniques and alternatives will be cultivated quickly if there is sufficient demand for the alternative fuels and biofuels. What could result from the shock to food prices is a very quick move toward some of the non-food biofuels as are listed above amongst the other policy alternatives.

Impact on Developing Countries

Demand for fuel in rich countries is now competing against demand for food in poor countries. The increase in world grain consumption in 2006 happened due to the increase in consumption for fuel, not human consumption. The grain required to fill a 25 US gallons (95 L) fuel tank with ethanol will feed one person for a year.

Several factors combine to make recent grain and oilseed price increases impact poor countries more:

- Poor people buy more grains (e.g. wheat), and are more exposed to grain price changes.
- Poor people spend a higher portion of their income on food, so increasing food prices influence them more.
- Aid organizations which buy food and send it to poor countries see more need when prices go up but are able to buy less food on the same budget.

The impact is not all negative. The Food and Agriculture Organization (FAO) recognizes the potential opportunities that the growing biofuel market offers to small farmers and aquaculturers around the world and has recommended small-scale financing to help farmers in poor countries produce local biofuel.

On the other hand, poor countries that do substantial farming have increased profits due to biofuels. If vegetable oil prices double, the profit margin could more than double. In the past rich countries have been dumping subsidized grains at below cost prices into poor countries and hurting the local farming industries. With biofuels using grains the rich countries no longer have grain surpluses to get rid of. Farming in poor countries is seeing healthier profit margins and expanding.

Interviews with local peasants in southern Ecuador provide strong anecdotal evidence that the high price of corn is encouraging the burning of tropical forests. The destruction of tropical forests now account for 20% of all greenhouse gas emmisons.

National Corn Growers Association

US government subsidies for making ethanol from corn have been attacked as the main cause of the food vs fuel problem. To defend themselves, the National Corn Growers Association has published their views on this issue. They consider the "food vs fuel" argument to be a fallacy that is "fraught with misguided logic, hyperbole and scare tactics."

Claims made by the NCGA include:

- Corn growers have been and will continue to produce enough corn so that supply and demand meet and there is no shortage. Farmers make their planting decisions based on signals from the marketplace. If demand for corn is high and projected revenue-per-acre is strong relative to other crops, farmers will plant more corn. In 2007 US farmers planted 92,900,000 acres (376,000 km^2) with corn, 19% more acres than they did in 2006.
- The U.S. has doubled corn yields over the last 40 years and expects to double them again in the next 20 years. With twice as much corn from each acre, corn can be put to new uses without taking food from the hungry or causing deforestation.

- US consumers buy things like corn flakes where the cost of the corn per box is around 5 cents. Most of the cost is packaging, advertising, shipping, etc. Only about 19% of the US retail food prices can be attributed to the actual cost of food inputs like grains and oilseeds. So if the price of a bushel of corn goes up, there may be no noticeable impact on US retail food prices. The US retail food price index has gone up only a few percent per year and is expected to continue to have very small increases.
- Most of the corn produced in the US is field corn, not sweet corn, and not digestible by humans in its raw form. Most corn is used for livestock feed and not human food, even the portion that is exported.
- Only the starch portion of corn kernels is converted to ethanol. The rest (protein, fat, vitamins and minerals) is passed through to the feed coproducts or human food ingredients.
- One of the most significant and immediate benefits of higher grain prices is a dramatic reduction in federal farm support payments. According to the U.S. Department of Agriculture, corn farmers received $8.8 billion in government support in 2006. Because of higher corn prices, payments are expected to drop to $2.1 billion in 2007, a 76 percent reduction.
- While the EROEI and economics of corn based ethanol are a bit weak, it paves the way for cellulosic ethanol which should have much better EROEI and economics.
- While basic nourishment is clearly important, fundamental societal needs of energy, mobility, and energy security are too. If farmers crops can help their country in these areas also, it seems right to do so.

Since reaching record high prices in June 2008, corn prices fell 50% by October 2008, declining sharply together with other commodities, including oil. As ethanol production from corn has continue at the same levels, some have argued that this trend shows the belief that the increased demand for corn to produce ethanol was mistaken. "Analysts, including some in the ethanol sector, say ethanol demand adds about 75 cents to $1.00 per bushel to the price of corn, as a rule of thumb. Other analysts say it adds around 20 percent, or just under 80 cents per bushel at current prices. Those estimates hint that $4 per bushel corn might be priced at only $3 without demand for ethanol fuel.". These industry sources consider that a speculative bubble in the commodity markets holding positions in corn futures was the main driver behind the observed hike in corn prices affecting food supply.

Controversy within the International System

The United States and Brazil lead the industrial world in global ethanol production, with Brazil as the world's largest exporter and biofuel industry leader. In 2006 the U.S. produced 18.4 billion liters (4.86 billion gallons), closely followed by Brazil with 16.3 billion liters (4.3 billion gallons), producing together 70% of the world's ethanol market and nearly 90% of ethanol used as fuel. These countries are followed by China with 7.5%, and India with 3.7% of the global market share.

Since 2007, the concerns, criticisms and controversy surrounding the food vs biofuels issue has reached the international system, mainly heads of states, and inter-governmental organizations (IGOs), such as the United Nations and several of its agencies, particularly the Food and Agriculture

Organization (FAO) and the World Food Programme (WFP); the International Monetary Fund; the World Bank; and agencies within the European Union.

The 2007 Controversy: Ethanol Diplomacy in the Americas

Presidents Luiz Inácio Lula da Silva and George W. Bush during Bush's visit to Brazil, March 2007.

In March 2007, "ethanol diplomacy" was the focus of President George W. Bush's Latin American tour, in which he and Brazil's president, Luiz Inácio Lula da Silva, were seeking to promote the production and use of sugar cane based ethanol throughout Latin America and the Caribbean. The two countries also agreed to share technology and set international standards for biofuels. The Brazilian sugar cane technology transfer will permit various Central American countries, such as Honduras, Nicaragua, Costa Rica and Panama, several Caribbean countries, and various Andean Countries tariff-free trade with the U.S. thanks to existing concessionary trade agreements. Even though the U.S. imposes a USD 0.54 tariff on every gallon of imported ethanol, the Caribbean nations and countries in the Central American Free Trade Agreement are exempt from such duties if they produce ethanol from crops grown in their own countries. The expectation is that using Brazilian technology for refining sugar cane based ethanol, such countries could become exporters to the United States in the short-term. In August 2007, Brazil's President toured Mexico and several countries in Central America and the Caribbean to promote Brazilian ethanol technology.

This alliance between the U.S. and Brazil generated some negative reactions. While Bush was in São Paulo as part of the 2007 Latin American tour, Venezuela's President Hugo Chavez, from Buenos Aires, dismissed the ethanol plan as "a crazy thing" and accused the U.S. of trying "to substitute the production of foodstuffs for animals and human beings with the production of foodstuffs for vehicles, to sustain the American way of life." Chavez' complaints were quickly followed by then Cuban President Fidel Castro, who wrote that "you will see how many people among the hungry masses of our planet will no longer consume corn." "Or even worse," he continued, "by offering financing to poor countries to produce ethanol from corn or any other kind of food, no tree will be left to defend humanity from climate change."' Daniel Ortega, Nicaragua's President, and one of the preferential recipients of Brazil technical aid, said that "we reject the gibberish of those who applaud Bush's totally absurd proposal, which attacks the food security rights of Latin Americans and Africans, who are major corn consumers", however, he voiced support for sugar cane based ethanol during Lula's visit to Nicaragua.

The 2008 Controversy: Global Food Prices

As a result of the international community's concerns regarding the steep increase in food prices, on April 14, 2008, Jean Ziegler, the United Nations Special Rapporteur on the Right to Food, at the Thirtieth Regional Conference of the Food and Agriculture Organization (FAO) in Brasília, called biofuels a "crime against humanity", a claim he had previously made in October 2007, when he called for a 5-year ban for the conversion of land for the production of biofuels. The previous day, at their Annual International Monetary Fund and World Bank Group meeting at Washington, D.C., the World Bank's President, Robert Zoellick, stated that "While many worry about filling their gas tanks, many others around the world are struggling to fill their stomachs. And it's getting more and more difficult every day."

Luiz Inácio Lula da Silva gave a strong rebuttal, calling both claims "fallacies resulting from commercial interests", and putting the blame instead on U.S. and European agricultural subsidies, and a problem restricted to U.S. ethanol produced from maize. He also said that "biofuels aren't the villain that threatens food security." In the middle of this new wave of criticism, Hugo Chavez reaffirmed his opposition and said that he is concerned that "so much U.S.-produced corn could be used to make biofuel, instead of feeding the world's poor", calling the U.S. initiative to boost ethanol production during a world food crisis a "crime."

German Chancellor Angela Merkel said the rise in food prices is due to poor agricultural policies and changing eating habits in developing nations, not biofuels as some critics claim. On the other hand, British Prime Minister Gordon Brown called for international action and said Britain had to be "selective" in supporting biofuels, and depending on the UK's assessment of biofuels' impact on world food prices, "we will also push for change in EU biofuels targets". Stavros Dimas, European Commissioner for the Environment said through a spokeswoman that "there is no question for now of suspending the target fixed for biofuels", though he acknowledged that the EU had underestimated problems caused by biofuels.

On April 29, 2008, U.S. President George W. Bush declared during a press conference that "85 percent of the world's food prices are caused by weather, increased demand and energy prices", and recognized that "15 percent has been caused by ethanol". He added that "the high price of gasoline is going to spur more investment in ethanol as an alternative to gasoline. And the truth of the matter is it's in our national interests that our farmers grow energy, as opposed to us purchasing energy from parts of the world that are unstable or may not like us." Regarding the effect of agricultural subsidies on rising food prices, Bush said that "Congress is considering a massive, bloated farm bill that would do little to solve the problem. The bill Congress is now considering would fail to eliminate subsidy payments to multi-millionaire farmers", he continued, "this is the right time to reform our nation's farm policies by reducing unnecessary subsidies".

Just a week before this new wave of international controversy began, U.N. Secretary General Ban Ki-moon had commented that several U.N. agencies were conducting a comprehensive review of the policy on biofuels, as the world food price crisis might trigger global instability. He said "We need to be concerned about the possibility of taking land or replacing arable land because of these biofuels", then he added "While I am very much conscious and aware of these problems, at the same time you need to constantly look at having creative sources of energy, including biofuels. Therefore, at this time, just criticising biofuel may not be a good solution. I would urge we need to address these issues in a comprehensive manner." Regarding Jean Ziegler's proposal for a five-year ban, the U.N. Secretary rejected that proposal.

A report released by Oxfam in June 2008 criticized biofuel policies of high-income countries as neither a solution to the climate crisis nor the oil crisis, while contributing to the food price crisis. The report concluded that from all biofuels available in the market, Brazilian sugarcane ethanol is not very effective, but it is the most favorable biofuel in the world in term of cost and greenhouse gas balance. The report discusses some existing problems and potential risks, and asks the Brazilian government for caution to avoid jeopardizing its environmental and social sustainability. The report also says that: "Rich countries spent up to $15 billion last year supporting biofuels while blocking cheaper Brazilian ethanol, which is far less damaging for global food security."

A World Bank research report published on July 2008 found that from June 2002 to June 2008 "biofuels and the related consequences of low grain stocks, large land use shifts, speculative activity and export bans" pushed prices up by 70 percent to 75 percent. The study found that higher oil prices and a weak dollar explain 25-30% of total price rise. The study said that "...large increases in biofuels production in the United States and Europe are the main reason behind the steep rise in global food prices" and also stated that "Brazil's sugar-based ethanol did not push food prices appreciably higher". The Renewable Fuels Association (RFA) published a rebuttal based on the version leaked before its formal release. The RFA critique considers that the analysis is highly subjective and that the author "estimates the impact of global food prices from the weak dollar and the direct and indirect effect of high petroleum prices and attributes everything else to biofuels."

An economic assessment by the OECD also published on July 2008 agrees with the World Bank report regarding the negative effects of subsidies and trade restrictions, but found that the impact of biofuels on food prices are much smaller. The OECD study is also critical of the limited reduction of GHG emissions achieved from biofuels produced in Europe and North America, concluding that the current biofuel support policies would reduce greenhouse gas emissions from transport fuel by no more than 0.8 percent by 2015, while Brazilian ethanol from sugar cane reduces greenhouse gas emissions by at least 80 percent compared to fossil fuels. The assessment calls on governments for more open markets in biofuels and feedstocks in order to improve efficiency and lower costs. The OECD study concluded that "...current biofuel support measures alone are estimated to increase average wheat prices by about 5 percent, maize by around 7 percent and vegetable oil by about 19 percent over the next 10 years."

Another World Bank research report published on July 2010 found their previous study may have overestimated the contribution of biofuel production, as the paper concluded that "the effect of biofuels on food prices has not been as large as originally thought, but that the use of commodities by financial investors (the so-called "financialization of commodities") may have been partly responsible for the 2007/08 spike."

Sustainable Biofuel

Biofuels, in the form of liquid fuels derived from plant materials, are entering the market, driven by factors such as oil price spikes and the need for increased energy security. However, many of the biofuels that are currently being supplied have been criticised for their adverse impacts on the natural environment, food security, and land use.

The challenge is to support biofuel development, including the development of new cellulosic technologies, with responsible policies and economic instruments to help ensure that biofuel commercialization is sustainable. Responsible commercialization of biofuels represents an opportunity to enhance sustainable economic prospects in Africa, Latin America and Asia.

Biofuels have a limited ability to replace fossil fuels and should not be regarded as a 'silver bullet' to deal with transport emissions. However, they offer the prospect of increased market competition and oil price moderation. A healthy supply of alternative energy sources will help to combat gasoline price spikes and reduce dependency on fossil fuels, especially in the transport sector. Using transportation fuels more efficiently is also an integral part of a sustainable transport strategy.

Biofuel Options

Biofuel development and use is a complex issue because there are many biofuel options which are available. Biofuels, such as ethanol and biodiesel, are currently produced from the products of conventional food crops such as the starch, sugar and oil feedstocks from crops that include wheat, maize, sugar cane, palm oil and oilseed rape. Some researchers fear that a major switch to biofuels from such crops would create a direct competition with their use for food and animal feed, and claim that in some parts of the world the economic consequences are already visible, other researchers look at the land available and the enormous areas of idle and abandoned land and claim that there is room for a large proportion of biofuel also from conventional crops.

Second generation biofuels are now being produced from a much broader range of feedstocks including the cellulose in dedicated energy crops (perennial grasses such as switchgrass and Miscanthus giganteus), forestry materials, the co-products from food production, and domestic vegetable waste. Advances in the conversion processes will improve the sustainability of biofuels, through better efficiencies and reduced environmental impact of producing biofuels, from both existing food crops and from cellulosic sources.

In 2007, Ronald Oxburgh suggested in *The Courier-Mail* that production of biofuels could be either responsible or irresponsible and had several trade-offs: "Produced responsibly they are a sustainable energy source that need not divert any land from growing food nor damage the environment; they can also help solve the problems of the waste generated by Western society; and they can create jobs for the poor where previously were none. Produced irresponsibly, they at best offer no climate benefit and, at worst, have detrimental social and environmental consequences. In other words, biofuels are pretty much like any other product. In 2008 the Nobel prize-winning chemist Paul J. Crutzen published findings that the release of nitrous oxide (N_2O) emissions in the production of biofuels means that they contribute more to global warming than the fossil fuels they replace.

According to the Rocky Mountain Institute, sound biofuel production practices would not hamper food and fibre production, nor cause water or environmental problems, and would enhance soil fertility. The selection of land on which to grow the feedstocks is a critical component of the ability of biofuels to deliver sustainable solutions. A key consideration is the minimisation of biofuel competition for prime cropland.

Plants Used as Sustainable Biofuel

Sugarcane in Brazil

Sugarcane (*Saccharum officinarum*) plantation ready for harvest, Ituverava, São Paulo State, Brazil.

Mechanized harvesting of sugarcane, Piracicaba, São Paulo, Brazil.

Cosan's Costa Pinto sugar cane mill and ethanol distillery plant at Piracicaba, São Paulo, Brazil.

Brazil's production of ethanol fuel from sugarcane dates back to the 1970s, as a governmental response to the 1973 oil crisis. Brazil is considered the biofuel industry leader and the world's first sustainable biofuels economy. In 2010 the U.S. Environmental Protection Agency designated Brazilian sugarcane ethanol as an advanced biofuel due to EPA's estimated 61% reduction of total life cycle greenhouse gas emissions, including direct indirect land use change emissions. Brazil sugarcane ethanol fuel program success and sustainability is based on the most efficient agricultural technology for sugarcane cultivation in the world, uses modern equipment and cheap sugar cane as feedstock, the residual cane-waste (bagasse) is used to process heat and power, which results in a very competitive price and also in a high energy balance (output energy/input energy), which varies from 8.3 for average conditions to 10.2 for best practice production.

A report commissioned by the United Nations, based on a detailed review of published research up to mid-2009 as well as the input of independent experts world-wide, found that ethanol from sugar cane as produced in Brazil "*in some circumstances does better than just "zero emission". If grown and processed correctly, it has negative emission, pulling CO2 out of the atmosphere, rather than adding it.* In contrast, the report found that U.S. use of maize for biofuel is less efficient, as sugarcane can lead to emissions reductions of between 70% and well over 100% when substituted for gasoline. Several other studies have shown that sugarcane-based ethanol reduces greenhouse gases by 86 to 90% if there is no significant land use change.

In another study commissioned by the Dutch government in 2006 to evaluate the sustainability of Brazilian bioethanol concluded that there is sufficient water to supply all foreseeable long-term water requirements for sugarcane and ethanol production. This evaluation also found that consumption of agrochemicals for sugar cane production is lower than in citric, corn, coffee and soybean cropping. The study found that development of resistant sugar cane varieties is a crucial aspect of disease and pest control and is one of the primary objectives of Brazil's cane genetic improvement programs. Disease control is one of the main reasons for the replacement of a commercial variety of sugar cane.

Another concern is the fact that sugarcane fields are traditionally burned just before harvest to avoid harm to the workers, by removing the sharp leaves and killing snakes and other harmful animals, and also to fertilize the fields with ash. Mechanization will reduce pollution from burning fields and has higher productivity than people, and due to mechanization the number of temporary workers in the sugarcane plantations has already declined. By the 2008 harvest season, around 47% of the cane was collected with harvesting machines.

Regarding the negative impacts of the potential direct and indirect effect of land use changes on carbon emissions, the study commissioned by the Dutch government concluded that "it is very difficult to determine the indirect effects of further land use for sugar cane production (i.e. sugar cane replacing another crop like soy or citrus crops, which in turn causes additional soy plantations replacing pastures, which in turn may cause deforestation), and also not logical to attribute all these soil carbon losses to sugar cane". The Brazilian agency Embrapa estimates that there is enough agricultural land available to increase at least 30 times the existing sugarcane plantation without endangering sensible ecosystems or taking land destined for food crops. Most future growth is expected to take place on abandoned pasture lands, as it has been the historical trend in São Paulo state. Also, productivity is expected to improve even further based on current biotechnology research, genetic improvement, and better agronomic practices, thus contributing to reduce land demand for future sugarcane cultures.

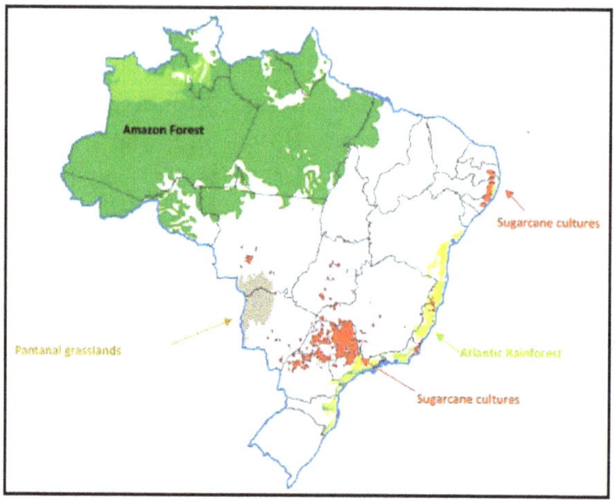

Location of environmentally valuable areas with respect to sugarcane plantations. São Paulo, located in the Southeast Region of Brazil, concentrates two-thirds of sugarcane cultures.

Another concern is the risk of clearing rain forests and other environmentally valuable land for sugarcane production, such as the Amazonia, the Pantanal or the Cerrado. Embrapa has rebutted this concern explaining that 99.7% of sugarcane plantations are located at least 2,000 km from the Amazonia, and expansion during the last 25 years took place in the Center-South region, also far away from the Amazonia, the Pantanal or the Atlantic forest. In São Paulo state growth took place in abandoned pasture lands. The impact assessment commissioned by the Dutch government supported this argument.

In order to guarantee a sustainable development of ethanol production, in September 2009 the government issued by decree a countrywide agroecological land use zoning to restrict sugarcane growth in or near environmentally sensitive areas. According to the new criteria, 92.5% of the Brazilian territory is not suitable for sugarcane plantation. The government considers that the suitable areas are more than enough to meet the future demand for ethanol and sugar in the domestic and international markets foreseen for the next decades.

Regarding the food vs fuel issue, a World Bank research report published on July 2008 found that *"Brazil's sugar-based ethanol did not push food prices appreciably higher"*. This research paper also concluded that Brazil's sugar cane–based ethanol has not raised sugar prices significantly. An economic assessment report also published in July 2008 by the OECD agrees with the World Bank report regarding the negative effects of subsidies and trade restrictions, but found that the impact of biofuels on food prices are much smaller. A study by the Brazilian research unit of the Fundação Getúlio Vargas regarding the effects of biofuels on grain prices concluded that the major driver behind the 2007-2008 rise in food prices was speculative activity on futures markets under conditions of increased demand in a market with low grain stocks. The study also concluded that there is no correlation between Brazilian sugarcane cultivated area and average grain prices, as on the contrary, the spread of sugarcane was accompanied by rapid growth of grain crops in the country.

Jatropha

India and Africa

Jatropha gossipifolia in Hyderabad, India.

Crops like Jatropha, used for biodiesel, can thrive on marginal agricultural land where many trees and crops won't grow, or would produce only slow growth yields. Jatropha cultivation provides benefits for local communities.

Cultivation and fruit picking by hand is labour-intensive and needs around one person per hectare. In parts of rural India and Africa this provides much-needed jobs - about 200,000 people worldwide now find employment through jatropha. Moreover, villagers often find that they can grow other crops in the shade of the trees. Their communities will avoid importing expensive diesel and there will be some for export too.

Cambodia

Cambodia has no proven fossil fuel reserves, and is almost completely dependent on imported diesel fuel for electricity production. Consequently, Cambodians face an insecure supply and pay some of the highest energy prices in the world. The impacts of this are widespread and may hinder economic development.

Biofuels may provide a substitute for diesel fuel that can be manufactured locally for a lower price, independent of the international oil price. The local production and use of biofuel also offers other benefits such as improved energy security, rural development opportunities and environmental benefits. The Jatropha curcas species appears to be a particularly suitable source of biofuel as it already grows commonly in Cambodia. Local sustainable production of biofuel in Cambodia, based on the Jatropha or other sources, offers good potential benefits for the investors, the economy, rural communities and the environment.

Mexico

Jatropha is native to Mexico and Central America and was likely transported to India and Africa in the 1500s by Portuguese sailors convinced it had medicinal uses. In 2008, recognizing the need to diversify its sources of energy and reduce emissions, Mexico passed a law to push developing biofuels that don't threaten food security and the agriculture ministry has since identified some 2.6 million hectares (6.4 million acres) of land with a high potential to produce jatropha. The Yucatán Peninsula, for instance, in addition to being a corn producing region, also contains abandoned sisal plantations, where the growing of Jatropha for biodiesel production would not displace food.

On April 1, 2011 Interjet completed the first Mexican aviation biofuels test flight on an Airbus A320. The fuel was a 70:30 traditional jet fuel biojet blend produced from Jatropha oil provided by three Mexican producers, Global Energías Renovables (a wholly owned subsidiary of U.S.-based Global Clean Energy Holdings, Bencafser S.A. and Energy JH S.A. Honeywell's UOP processed the oil into Bio-SPK (Synthetic Paraffinic Kerosene) . Global Energías Renovables operates the largest Jatropha farm in the Americas.

On August 1, 2011 Aeromexico, Boeing, and the Mexican Government participated in the first bio-jet powered transcontinental flight in aviation history. The flight from Mexico City to Madrid used a blend of 70 percent traditional fuel and 30 percent biofuel (aviation biofuel). The biojet was produced entirely from Jatropha oil.

Pongamia Pinnata in Australia and India

Pongamia pinnata seeds in Brisbane, Australia.

Pongamia pinnata is a legume native to Australia, India, Florida (USA) and most tropical regions, and is now being invested in as an alternative to Jatropha for areas such as Northern Australia, where Jatropha is classed as a noxious weed. Commonly known as simply 'Pongamia', this tree is currently being commercialised in Australia by Pacific Renewable Energy, for use as a Diesel replacement for running in modified Diesel engines or for conversion to Biodiesel using 1st or 2nd Generation Biodiesel techniques, for running in unmodified Diesel engines.

Sweet Sorghum in India

Sweet sorghum overcomes many of the shortcomings of other biofuel crops. With sweet sorghum, only the stalks are used for biofuel production, while the grain is saved for food or livestock feed. It is not in high demand in the global food market, and thus has little impact on food prices and food security. Sweet sorghum is grown on already-farmed drylands that are low in carbon storage capacity, so concerns about the clearing of rainforest do not apply. Sweet sorghum is easier and cheaper to grow than other biofuel crops in India and does not require irrigation, an important consideration in dry areas. Some of the Indian sweet sorghum varieties are now grown in Uganda for ethanol production.

A study by researchers at the International Crops Research Institute for the Semi-Arid Tropics (ICRISAT) found that growing sweet sorghum instead of grain sorghum could increase farmers incomes by US$40 per hectare per crop because it can provide food, feed and fuel. With grain sorghum currently grown on over 11 million hectares (ha) in Asia and on 23.4 million ha in Africa, a switch to sweet sorghum could have a considerable economic impact.

International Collaboration on Sustainable Biofuels

Roundtable on Sustainable Biofuels

Public attitudes and the actions of key stakeholders can play a crucial role in realising the potential of sustainable biofuels. Informed discussion and dialogue, based both on scientific research and an understanding of public and stakeholder views, is important.

The Roundtable on Sustainable Biofuels is an international initiative which brings together farmers, companies, governments, non-governmental organizations, and scientists who are interested in the sustainability of biofuels production and distribution. During 2008, the Roundtable used meetings, teleconferences, and online discussions to develop a series of principles and criteria for sustainable biofuels production.

In 2008, the Roundtable for Sustainable Biofuels released its proposed standards for sustainable biofuels. This includes 12 principles:

1. "Biofuel production shall follow international treaties and national laws regarding such things as air quality, water resources, agricultural practices, labor conditions, and more.
2. Biofuels projects shall be designed and operated in participatory processes that involve all relevant stakeholders in planning and monitoring.
3. Biofuels shall significantly reduce greenhouse gas emissions as compared to fossil fuels. The principle seeks to establish a standard methodology for comparing greenhouse gases (GHG) benefits.
4. Biofuel production shall not violate human rights or labor rights, and shall ensure decent work and the well-being of workers.
5. Biofuel production shall contribute to the social and economic development of local, rural and indigenous peoples and communities.
6. Biofuel production shall not impair food security.
7. Biofuel production shall avoid negative impacts on biodiversity, ecosystems and areas of high conservation value.
8. Biofuel production shall promote practices that improve soil health and minimize degradation.

9. Surface and groundwater use will be optimized and contamination or depletion of water resources minimized.
10. Air pollution shall be minimized along the supply chain.
11. Biofuels shall be produced in the most cost-effective way, with a commitment to improve production efficiency and social and environmental performance in all stages of the biofuel value chain.
12. Biofuel production shall not violate land rights".

In April 2011, the Roundtable on Sustainable Biofuels launched a set of comprehensive sustainability criteria - the "RSB Certification System." Biofuels producers that meet to these criteria are able to show buyers and regulators that their product has been obtained without harming the environment or violating human rights.

Sustainable Biofuels Consensus

The Sustainable Biofuels Consensus is an international initiative which calls upon governments, the private sector, and other stakeholders to take decisive action to ensure the sustainable trade, production, and use of biofuels. In this way biofuels may play a key role in energy sector transformation, climate stabilization, and resulting worldwide revitalisation of rural areas.

The Sustainable Biofuels Consensus envisions a "landscape that provides food, fodder, fiber, and energy, which offers opportunities for rural development; that diversifies energy supply, restores ecosystems, protects biodiversity, and sequesters carbon".

Better Sugarcane Initiative / Bonsucro

In 2008, a multi-stakeholder process was initiated by the World Wildlife Fund and the International Finance Corporation, the private development arm of the World Bank, bringing together industry, supply chain intermediaries, end-users, farmers and civil society organisations to develop standards for certifying the derivative products of sugar cane, one of which is ethanol fuel.

The Bonsucro standard is based around a definition of sustainability which is founded on five principles:

1. Obey the law.
2. Respect human rights and labour standards.
3. Manage input, production and processing efficiencies to enhance sustainability.
4. Actively manage biodiversity and ecosystem services.
5. Continuously improve key areas of the business.

Biofuel producers that wish to sell products marked with the Bonsucro standard must both ensure that they product to the Production Standard, and that their downstream buyers meet the Chain of Custody Standard. In addition, if they wish to sell to the European market and count against the EU Renewable Energy Directive, then they must adhere to the Bonsucro EU standard, which includes specific greenhouse gas calculations following European Commission calculation guidelines.

Sustainability Standards

Several countries and regions have introduced policies or adopted standards to promote sustainable biofuels production and use, most prominently the European Union and the United States. The 2009 EU Renewable Energy Directive, which requires 10 percent of transportation energy from renewable energy by 2020, is the most comprehensive mandatory sustainability standard in place as of 2010. The Directive requires that the lifecycle greenhouse gas emissions of biofuels consumed be at least 50 percent less than the equivalent emissions from gasoline or diesel by 2017 (and 35 percent less starting in 2011). Also, the feedstocks for biofuels "should not be harvested from lands with high biodiversity value, from carbon-rich or forested land, or from wetlands".

As with the EU, the U.S. Renewable Fuel Standard (RFS) and the California Low Carbon Fuel Standard (LCFS) both require specific levels of lifecycle greenhouse gas reductions compared to equivalent fossil fuel consumption. The RFS requires that at least half of the biofuels production mandated by 2022 should reduce lifecycle emissions by 50 percent. The LCFS is a performance standard that calls for a minimum of 10 percent emissions reduction per unit of transport energy by 2020. Both the U.S. and California standards currently address only greenhouse gas emissions, but California plans to "expand its policy to address other sustainability issues associated with liquid biofuels in the future".

In 2009, Brazil also adopted new sustainability policies for sugarcane ethanol, including "zoning regulation of sugarcane expansion and social protocols".

Oil Price Moderation

Biofuels offer the prospect of real market competition and oil price moderation. According to the Wall Street Journal, crude oil would be trading 15 per cent higher and gasoline would be as much as 25 per cent more expensive, if it were not for biofuels. A healthy supply of alternative energy sources will help to combat gasoline price spikes.

Sustainable Transport

Biofuels have a limited ability to replace fossil fuels and should not be regarded as a 'silver bullet' to deal with transport emissions. Biofuels on their own cannot deliver a sustainable transport system and so must be developed as part of an integrated approach, which promotes other renewable energy options and energy efficiency, as well as reducing the overall energy demand and need for transport. Consideration needs to be given to the development of hybrid and fuel cell vehicles, public transport, and better town and rural planning.

In December 2008 an Air New Zealand jet completed the world's first commercial aviation test flight partially using jatropha-based fuel. More than a dozen performance tests were undertaken in the two-hour test flight which departed from Auckland International Airport. A biofuel blend of 50:50 jatropha and Jet A1 fuel was used to power one of the Boeing 747-400's Rolls-Royce RB211 engines. Air New Zealand set several criteria for its jatropha, requiring that "the land it came from was neither forest nor virgin grassland in the previous 20 years, that the soil and climate it came from is not suitable for the majority of food crops and that the farms are rain fed and not mechanically irrigated". The company has also set general sustainability criteria, saying that such biofuels must not compete with food resources, that they must be as good as traditional jet fuels, and that they should be cost competitive.

In January 2009, Continental Airlines used a sustainable biofuel to power a commercial aircraft for the first time in North America. This demonstration flight marks the first sustainable biofuel demonstration flight by a commercial carrier using a twin-engined aircraft, a Boeing 737-800, powered by CFM International CFM56-7B engines. The biofuel blend included components derived from algae and jatropha plants. The algae oil was provided by Sapphire Energy, and the jatropha oil by Terasol Energy.

In March 2011, Yale University research showed significant potential for sustainable aviation fuel based on jatropha-curcas. According to the research, if cultivated properly, "jatropha can deliver many benefits in Latin America and greenhouse gas reductions of up to 60 percent when compared to petroleum-based jet fuel". Actual farming conditions in Latin America were assessed using sustainability criteria developed by the Roundtable on Sustainable Biofuels. Unlike previous research, which used theoretical inputs, the Yale team conducted many interviews with jatropha farmers and used "field measurements to develop the first comprehensive sustainability analysis of actual projects".

As of June 2011, revised international aviation fuel standards officially allow commercial airlines to blend conventional jet fuel with up to 50 percent biofuels. The renewable fuels "can be blended with conventional commercial and military jet fuel through requirements in the newly issued edition of ASTM D7566, Specification for Aviation Turbine Fuel Containing Synthesized Hydrocarbons".

In December 2011, the FAA awarded $7.7 million to eight companies to advance the development of commercial aviation biofuels, with a special focus on alcohol to jet fuel. The FAA is assisting in the development of a sustainable fuel (from alcohols, sugars, biomass, and organic matter such as pyrolysis oils) that can be "dropped in" to aircraft without changing current practices and infrastructure. The research will test how the new fuels affect engine durability and quality control standards.

GreenSky London, a biofuels plant under construction in 2014, will take in some 500,000 tonnes of municipal rubbish and change the organic component into 60,000 tonnes of jet fuel, and 40 megawatts of power. By the end of 2015, all British Airways flights from London City Airport will be fuelled by waste and rubbish discarded by London residents.

Environmental Impact of Biodiesel

The environmental impact of biodiesel is diverse.

Greenhouse Gas Emissions

An often mentioned incentive for using biodiesel is its capacity to lower greenhouse gas emissions compared to those of fossil fuels. Whether this is true or not depends on many factors. Especially the effects from land use change have potential to cause even more emissions than what would be caused by using fossil fuels alone.

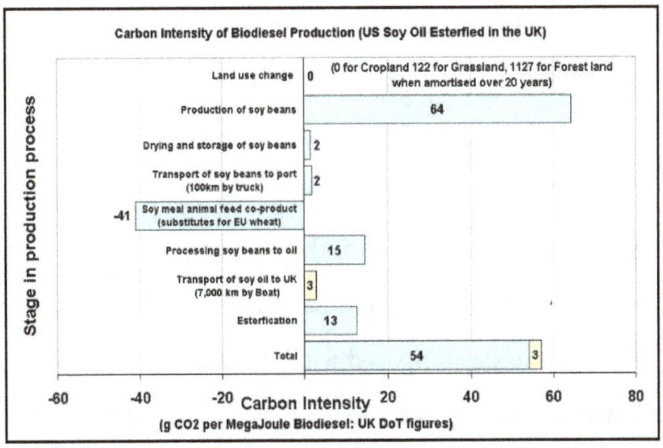

Calculation of Carbon Intensity of Soy biodiesel grown in the US and burnt in the UK, using figures calculated by the UK government for the purposes of the Renewable transport fuel obligation.

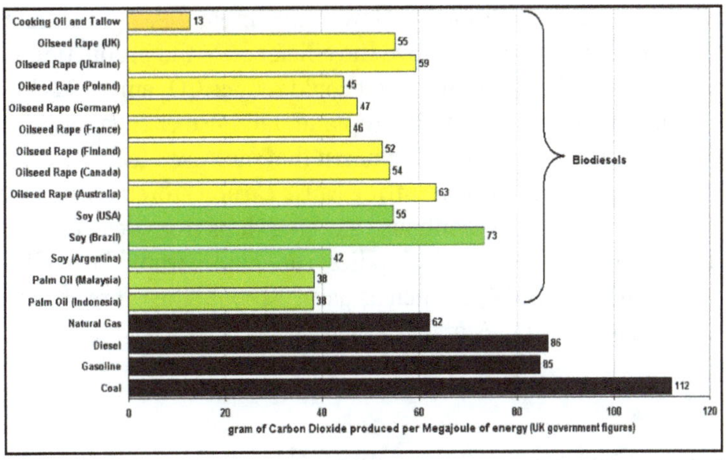

Graph of UK figures for the Carbon Intensity of Biodiesels and fossil fuels. This graph assumes that all biodiesel is used in its country of origin. It also assumes that the diesel is produced from pre-existing croplands rather than by changing land use.

Carbon dioxide is one of the major greenhouse gases. Although the burning of biodiesel produces carbon dioxide emissions similar to those from ordinary fossil fuels, the plant feedstock used in the production absorbs carbon dioxide from the atmosphere when it grows. Plants absorb carbon dioxide through a process known as photosynthesis which allows it to store energy from sunlight in the form of sugars and starches. After the biomass is converted into biodiesel and burned as fuel the energy and carbon is released again. Some of that energy can be used to power an engine while the carbon dioxide is released back into the atmosphere.

When considering the total amount of greenhouse gas emissions it is therefore important to consider the whole production process and what indirect effects such production might cause. The effect on carbon dioxide emissions is highly dependent on production methods and the type of feedstock used. Calculating the carbon intensity of biofuels is a complex and inexact process, and is highly dependent on the assumptions made in the calculation. A calculation usually includes;

- Emissions from growing the feedstock (e.g. Petrochemicals used in fertilizers).
- Emissions from transporting the feedstock to the factory.
- Emissions from processing the feedstock into biodiesel.

Other factors can be very significant but are sometimes not considered. These include:

- Emissions from the change in land use of the area where the fuel feedstock is grown.
- Emissions from transportation of the biodiesel from the factory to its point of use.
- The efficiency of the biodiesel compared with standard diesel.
- The amount of Carbon Dioxide produced at the tail pipe. (Biodiesel can produce 4.7% more).
- The benefits due to the production of useful by-products, such as cattle feed or glycerine.

If land use change is not considered and assuming today's production methods, biodiesel from rapeseed and sunflower oil produce 45%-65% lower greenhouse gas emissions than petrodiesel. However, there is ongoing research to improve the efficiency of the production process. Biodiesel produced from used cooking oil or other waste fat could reduce CO_2 emissions by as much as 85%. As long as the feedstock is grown on existing cropland, land use change has little or no effect on greenhouse gas emissions. However, there is concern that increased feedstock production directly affects the rate of deforestation. Such clearcutting cause carbon stored in the forest, soil and peat layers to be released. The amount of greenhouse gas emissions from deforestation is so large that the benefits from lower emissions (caused by biodiesel use alone) would be negligible for hundreds of years. Biofuel produced from feedstock such as palm oil could therefore cause much higher carbon dioxide emissions than some types of fossil fuels.

Pollution

In the United States, biodiesel is the only alternative fuel to have successfully completed the Health Effects Testing requirements (Tier I and Tier II) of the Clean Air Act (1990).

Biodiesel can reduce the direct tailpipe-emission of particulates, small particles of solid combustion products, on vehicles with particulate filters by as much as 20 percent compared with low-sulfur (< 50 ppm) diesel. Particulate emissions as the result of production are reduced by around 50 percent compared with fossil-sourced diesel. (Beer et al., 2004). Biodiesel has a higher cetane rating than petrodiesel, which can improve performance and clean up emissions compared to crude petro-diesel (with cetane lower than 40). Biodiesel contains fewer aromatic hydrocarbons: benzofluoranthene: 56% reduction; Benzopyrenes: 71% reduction.

Biodegradation

A University of Idaho study compared biodegradation rates of biodiesel, neat vegetable oils, biodiesel and petroleum diesel blends, and neat 2-D diesel fuel. Using low concentrations of the product to be degraded (10 ppm) in nutrient and sewage sludge amended solutions, they demonstrated that biodiesel degraded at the same rate as a dextrose control and 5 times as quickly as petroleum diesel over a period of 28 days, and that biodiesel blends doubled the rate of petroleum diesel degradation through co-metabolism. The same study examined soil degradation using 10 000 ppm of biodiesel and petroleum diesel, and found biodiesel degraded at twice the rate of petroleum diesel in soil. In all cases, it was determined biodiesel also degraded more completely than petroleum diesel, which produced poorly degradable undetermined intermediates. Toxicity studies for the same project demonstrated no mortalities and few toxic effects on rats and rabbits with up to 5000 mg/kg of biodiesel. Petroleum diesel showed no mortalities at the same concentration either, however toxic effects such as hair loss and urinary discolouring were noted with concentrations of >2000 mg/l in rabbits.

Biodegradation in Aquatic Environments

As biodiesel becomes more widely used, it is important to consider how consumption affects water quality and aquatic ecosystems. Research examining the biodegradability of different biodiesel fuels found that all of the biofuels studied (including Neat Rapeseed oil, Neat Soybean oil, and their modified ester products) were "readily biodegradable" compounds, and had a relatively high biodegradation rate in water. Additionally, the presence of biodiesel can increase the rate of diesel biodegradation via co-metabolism. As the ratio of biodiesel is increased in biodiesel/diesel mixtures, the faster the diesel is degraded. Another study using controlled experimental conditions also showed that fatty acid methyl esters, the primary molecules in biodiesel, degraded much faster than petroleum diesel in sea water.

Carbonyl Emissions

When considering the emissions from fossil fuel and biofuel use, research typically focuses on major pollutants such as hydrocarbons. It is generally recognized that using biodiesel in place of diesel results in a substantial reduction in regulated gas emissions, but there has been a lack of information in research literature about the non-regulated compounds which also play a role in air pollution. One study focused on the emissions of non-criteria carbonyl compounds from the burning of pure diesel and biodiesel blends in heavy-duty diesel engines. The results found that carbonyl emissions of formaldehyde, acetaldehyde, acrolein, acetone, propionaldehyde and butyraldehyde, were higher in biodiesel mixtures than emissions from pure diesel. Biodiesel use results in higher carbonyl emissions but lower total hydrocarbon emissions, which may be better as an alternative fuel source. Other studies have been done which conflict with these results, but comparisons are difficult to make due to various factors that differ between studies (such as types of fuel and engines used). In a paper which compared 12 research articles on carbonyl emissions from biodiesel fuel use, it found that 8 of the papers reported increased carbonyl compound emissions while 4 showed the opposite. This is evidence that there is still much research required on these compounds.

Deforestation

Deforestation, clearance or clearing is the removal of a forest or stand of trees where the land is thereafter converted to a non-forest use. Examples of deforestation include conversion of forestland to farms, ranches, or urban use. Tropical rainforests is where the most concentrated deforestation occurs. About 30% of Earth's land surface is covered by forests.

Satellite photograph of deforestation in progress in the Tierras Bajas project in eastern Bolivia.

In temperate mesic climates, natural regeneration of forest stands often will not occur in the absence of disturbance, whether natural or anthropogenic. Furthermore, biodiversity after regeneration harvest often mimics that found after natural disturbance, including biodiversity loss after naturally occurring rainforest destruction.

Deforestation occurs for multiple reasons: trees are cut down to be used or sold as fuel (sometimes in the form of charcoal) or timber, while cleared land is used as pasture for livestock and plantation. The removal of trees without sufficient reforestation has resulted in damage to habitat, biodiversity loss and aridity. It has adverse impacts on biosequestration of atmospheric carbon dioxide. Deforestation has also been used in war to deprive the enemy of cover for its forces and also vital resources. Modern examples of this were the use of Agent Orange by the British military in Malaya during the Malayan Emergency and the United States military in Vietnam during the Vietnam War. As of 2005, net deforestation rates have ceased to increase in countries with a per capita GDP of at least US$4,600. Deforested regions typically incur significant adverse soil erosion and frequently degrade into wasteland.

Disregard of ascribed value, lax forest management and deficient environmental laws are some of the factors that allow deforestation to occur on a large scale. In many countries, deforestation, both naturally occurring and human induced, is an ongoing issue. Deforestation causes extinction, changes to climatic conditions, desertification, and displacement of populations as observed by current conditions and in the past through the fossil record. More than half of all plant and land animal species in the world live in tropical forests.

Between 2000 and 2012, 2.3 million square kilometres (890,000 square miles) of forests around the earth were cut down. As a result of deforestation, only 6.2 million square kilometres (2.4 million square miles) remain of the original 16 million square kilometres (6 million square miles) of forest that formerly covered the earth.

Causes

According to the United Nations Framework Convention on Climate Change (UNFCCC) secretariat, the overwhelming direct cause of deforestation is agriculture. Subsistence farming is responsible for 48% of deforestation; commercial agriculture is responsible for 32% of deforestation; logging is responsible for 14% of deforestation and fuel wood removals make up 5% of deforestation.

Experts do not agree on whether industrial logging is an important contributor to global deforestation. Some argue that poor people are more likely to clear forest because they have no alternatives, others that the poor lack the ability to pay for the materials and labour needed to clear forest. One study found that population increases due to high fertility rates were a primary driver of tropical deforestation in only 8% of cases.

Other causes of contemporary deforestation may include corruption of government institutions, the inequitable distribution of wealth and power, population growth and overpopulation, and urbanization. Globalization is often viewed as another root cause of deforestation, though there are cases in which the impacts of globalization (new flows of labor, capital, commodities, and ideas) have promoted localized forest recovery.

The last batch of sawnwood from the peat forest in Indragiri Hulu, Sumatra, Indonesia. Deforestation for oil palm plantation.

In 2000 the United Nations Food and Agriculture Organization (FAO) found that "the role of population dynamics in a local setting may vary from decisive to negligible," and that deforestation can result from "a combination of population pressure and stagnating economic, social and technological conditions."

The degradation of forest ecosystems has also been traced to economic incentives that make forest conversion appear more profitable than forest conservation. Many important forest functions have no markets, and hence, no economic value that is readily apparent to the forests' owners or the communities that rely on forests for their well-being. From the perspective of the developing world, the benefits of forest as carbon sinks or biodiversity reserves go primarily to richer developed nations and there is insufficient compensation for these services. Developing countries feel that some countries in the developed world, such as the United States of America, cut down their forests centuries ago and benefited greatly from this deforestation, and that it is hypocritical to deny developing countries the same opportunities: that the poor shouldn't have to bear the cost of preservation when the rich created the problem.

Some commentators have noted a shift in the drivers of deforestation over the past 30 years. Whereas deforestation was primarily driven by subsistence activities and government-sponsored development projects like transmigration in countries like Indonesia and colonization in Latin

America, India, Java, and so on, during late 19th century and the earlier half of the 20th century. By the 1990s the majority of deforestation was caused by industrial factors, including extractive industries, large-scale cattle ranching, and extensive agriculture.

Environmental Problems

Atmospheric

Illegal slash and burn practice in Madagascar, 2010.

Deforestation is ongoing and is shaping climate and geography.

Deforestation is a contributor to global warming, and is often cited as one of the major causes of the enhanced greenhouse effect. Tropical deforestation is responsible for approximately 20% of world greenhouse gas emissions. According to the Intergovernmental Panel on Climate Change deforestation, mainly in tropical areas, could account for up to one-third of total anthropogenic carbon dioxide emissions. But recent calculations suggest that carbon dioxide emissions from deforestation and forest degradation (excluding peatland emissions) contribute about 12% of total anthropogenic carbon dioxide emissions with a range from 6 to 17%. Deforestation causes carbon dioxide to linger in the atmosphere. As carbon dioxide accrues, it produces a layer in the atmosphere that traps radiation from the sun. The radiation converts to heat which causes global warming, which is better known as the greenhouse effect. Plants remove carbon in the form of carbon dioxide from the atmosphere during the process of photosynthesis, but release some carbon dioxide back into the atmosphere during normal respiration. Only when actively growing can a tree or forest remove carbon, by storing it in plant tissues. Both the decay and burning of wood releases much of this stored carbon back to the atmosphere. In order for forests to take up carbon, there must be a net accumulation of wood. One way is for the wood to be harvested and turned into long-lived products, with new young trees replacing them. Deforestation may also cause carbon stores held in soil to be released. Forests can be either sinks or sources depending upon environmental circumstances. Mature forests alternate between being net sinks and net sources of carbon dioxide.

In deforested areas, the land heats up faster and reaches a higher temperature, leading to localized upward motions that enhance the formation of clouds and ultimately produce more rainfall. However, according to the Geophysical Fluid Dynamics Laboratory, the models used to investigate remote responses to tropical deforestation showed a broad but mild temperature increase all through the tropical atmosphere. The model predicted <0.2 °C warming for upper air at 700 mb and 500 mb. However, the model shows no significant changes in other areas besides the Tropics. Though the model showed no significant changes to the climate in areas other than the Tropics, this may not be the case since the model has possible errors and the results are never absolutely definite.

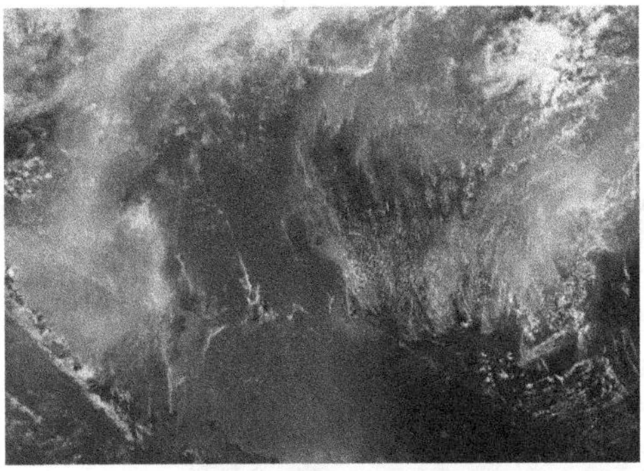

Fires on Borneo and Sumatra, 2006. People use slash-and-burn deforestation to clear land for agriculture.

Reducing emissions from deforestation and forest degradation (REDD) in developing countries has emerged as a new potential to complement ongoing climate policies. The idea consists in providing financial compensations for the reduction of greenhouse gas (GHG) emissions from deforestation and forest degradation".

Rainforests are widely believed by laymen to contribute a significant amount of the world's oxygen, although it is now accepted by scientists that rainforests contribute little net oxygen to the atmosphere and deforestation has only a minor effect on atmospheric oxygen levels. However, the incineration and burning of forest plants to clear land releases large amounts of CO_2, which contributes to global warming. Scientists also state that tropical deforestation releases 1.5 billion tons of carbon each year into the atmosphere.

Hydrological

The water cycle is also affected by deforestation. Trees extract groundwater through their roots and release it into the atmosphere. When part of a forest is removed, the trees no longer transpire this water, resulting in a much drier climate. Deforestation reduces the content of water in the soil and groundwater as well as atmospheric moisture. The dry soil leads to lower water intake for the trees to extract. Deforestation reduces soil cohesion, so that erosion, flooding and landslides ensue.

Shrinking forest cover lessens the landscape's capacity to intercept, retain and transpire precipitation. Instead of trapping precipitation, which then percolates to groundwater systems, deforested areas become sources of surface water runoff, which moves much faster than subsurface flows. That quicker transport of surface water can translate into flash flooding and more localized floods than would occur with the forest cover. Deforestation also contributes to decreased evapotranspiration, which lessens atmospheric moisture which in some cases affects precipitation levels downwind from the deforested area, as water is not recycled to downwind forests, but is lost in runoff and returns directly to the oceans. According to one study, in deforested north and northwest China, the average annual precipitation decreased by one third between the 1950s and the 1980s.

Trees, and plants in general, affect the water cycle significantly:

- their canopies intercept a proportion of precipitation, which is then evaporated back to the atmosphere (canopy interception);
- their litter, stems and trunks slow down surface runoff;
- their roots create macropores – large conduits – in the soil that increase infiltration of water;
- they contribute to terrestrial evaporation and reduce soil moisture via transpiration;
- their litter and other organic residue change soil properties that affect the capacity of soil to store water;
- their leaves control the humidity of the atmosphere by transpiring. 99% of the water absorbed by the roots moves up to the leaves and is transpired.

As a result, the presence or absence of trees can change the quantity of water on the surface, in the soil or groundwater, or in the atmosphere. This in turn changes erosion rates and the availability of water for either ecosystem functions or human services.

The forest may have little impact on flooding in the case of large rainfall events, which overwhelm the storage capacity of forest soil if the soils are at or close to saturation.

Tropical rainforests produce about 30% of our planet's fresh water.

Soil

Deforestation for the use of clay in the Brazilian city of Rio de Janeiro.
The hill depicted is Morro da Covanca, in Jacarepaguá.

Undisturbed forests have a very low rate of soil loss (erosion), approximately 2 metric tons per square kilometer (6 short tons per square mile). Deforestation generally increases rates of soil loss, by increasing the amount of runoff and reducing the protection of the soil from tree litter. This can be an advantage in excessively leached tropical rain forest soils. Forestry operations themselves also increase erosion through the development of (forest) roads and the use of mechanized equipment.

China's Loess Plateau was cleared of forest millennia ago. Since then it has been eroding, creating dramatic incised valleys, and providing the sediment that gives the Yellow River its yellow color and that causes the flooding of the river in the lower reaches (hence the river's nickname 'China's sorrow').

Removal of trees does not always increase erosion rates. In certain regions of southwest US, shrubs and trees have been encroaching on grassland. The trees themselves enhance the loss of grass between tree canopies. The bare intercanopy areas become highly erodible. The US Forest Service, in Bandelier National Monument for example, is studying how to restore the former ecosystem, and reduce erosion, by removing the trees.

Tree roots bind soil together, and if the soil is sufficiently shallow they act to keep the soil in place by also binding with underlying bedrock. Tree removal on steep slopes with shallow soil thus increases the risk of landslides, which can threaten people living nearby.

Biodiversity

Deforestation on a human scale results in decline in biodiversity, and on a natural global scale is known to cause the extinction of many species. The removal or destruction of areas of forest cover has resulted in a degraded environment with reduced biodiversity. Forests support biodiversity, providing habitat for wildlife; moreover, forests foster medicinal conservation. With forest biotopes being irreplaceable source of new drugs (such as taxol), deforestation can destroy genetic variations (such as crop resistance) irretrievably.

Illegal logging in Madagascar. In 2009, the vast majority of the illegally obtained rosewood was exported to China.

Since the tropical rainforests are the most diverse ecosystems on Earth and about 80% of the world's known biodiversity could be found in tropical rainforests, removal or destruction of significant areas of forest cover has resulted in a degraded environment with reduced biodiversity. A study in Rondônia, Brazil, has shown that deforestation also removes the microbial community which is involved in the recycling of nutrients, the production of clean water and the removal of pollutants.

It has been estimated that we are losing 137 plant, animal and insect species every single day due to rainforest deforestation, which equates to 50,000 species a year. Others state that tropical rainforest deforestation is contributing to the ongoing Holocene mass extinction. The known extinction rates from deforestation rates are very low, approximately 1 species per year from mammals and birds which extrapolates to approximately 23,000 species per year for all species. Predictions have been made that more than 40% of the animal and plant species in Southeast Asia could be wiped out in the 21st century. Such predictions were called into question by 1995 data that show that within regions of Southeast Asia much of the original forest has been converted to monospecific plantations, but that potentially endangered species are few and tree flora remains widespread and stable.

Scientific understanding of the process of extinction is insufficient to accurately make predictions about the impact of deforestation on biodiversity. Most predictions of forestry related biodiversity loss are based on species-area models, with an underlying assumption that as the forest declines species diversity will decline similarly. However, many such models have been proven to be wrong and loss of habitat does not necessarily lead to large scale loss of species. Species-area models are known to overpredict the number of species known to be threatened in areas where actual deforestation is ongoing, and greatly overpredict the number of threatened species that are widespread.

A recent study of the Brazilian Amazon predicts that despite a lack of extinctions thus far, up to 90 percent of predicted extinctions will finally occur in the next 40 years.

Economic Impact

Damage to forests and other aspects of nature could halve living standards for the world's poor and reduce global GDP by about 7% by 2050, a report concluded at the Convention on Biological Diversity (CBD) meeting in Bonn. Historically, utilization of forest products, including timber and fuel wood, has played a key role in human societies, comparable to the roles of water and cultivable land. Today, developed countries continue to utilize timber for building houses, and wood pulp for paper. In developing countries almost three billion people rely on wood for heating and cooking.

The forest products industry is a large part of the economy in both developed and developing countries. Short-term economic gains made by conversion of forest to agriculture, or over-exploitation of wood products, typically leads to loss of long-term income and long-term biological productivity. West Africa, Madagascar, Southeast Asia and many other regions have experienced lower revenue because of declining timber harvests. Illegal logging causes billions of dollars of losses to national economies annually.

The new procedures to get amounts of wood are causing more harm to the economy and overpower the amount of money spent by people employed in logging. According to a study, "in most areas studied, the various ventures that prompted deforestation rarely generated more than US$5 for every ton of carbon they released and frequently returned far less than US$1". The price on the European market for an offset tied to a one-ton reduction in carbon is 23 euro (about US$35).

Rapidly growing economies also have an effect on deforestation. Most pressure will come from the world's developing countries, which have the fastest-growing populations and most rapid economic (industrial) growth. In 1995, economic growth in developing countries reached nearly 6%, compared with the 2% growth rate for developed countries." As our human population grows, new homes, communities, and expansions of cities will occur. Connecting all of the new expansions will be roads, a very important part in our daily life. Rural roads promote economic development but also facilitate deforestation. About 90% of the deforestation has occurred within 100 km of roads in most parts of the Amazon.

The European Union is one of the largest importer of products made from illegal deforestation.

Forest Transition Theory

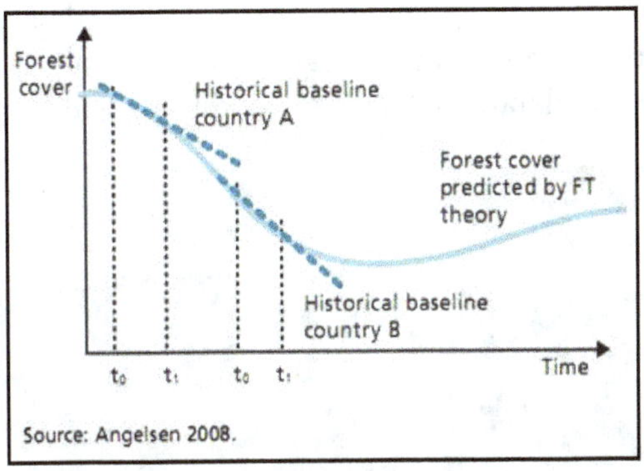

The forest transition and historical baselines.

The forest area change may follow a pattern suggested by the forest transition (FT) theory, whereby at early stages in its development a country is characterized by high forest cover and low deforestation rates (HFLD countries).

Then deforestation rates accelerate (HFHD, high forest cover – high deforestation rate), and forest cover is reduced (LFHD, low forest cover – high deforestation rate), before the deforestation rate slows (LFLD, low forest cover – low deforestation rate), after which forest cover stabilizes and eventually starts recovering. FT is not a "law of nature," and the pattern is influenced by national context (for example, human population density, stage of development, structure of the economy), global economic forces, and government policies. A country may reach very low levels of forest cover before it stabilizes, or it might through good policies be able to "bridge" the forest transition.

FT depicts a broad trend, and an extrapolation of historical rates therefore tends to underestimate future BAU deforestation for counties at the early stages in the transition (HFLD), while it tends to overestimate BAU deforestation for countries at the later stages (LFHD and LFLD).

Countries with high forest cover can be expected to be at early stages of the FT. GDP per capita captures the stage in a country's economic development, which is linked to the pattern of natural resource use, including forests. The choice of forest cover and GDP per capita also fits well with the two key scenarios in the FT:

(i) a forest scarcity path, where forest scarcity triggers forces (for example, higher prices of forest products) that lead to forest cover stabilization; and

(ii) an economic development path, where new and better off-farm employment opportunities associated with economic growth (= increasing GDP per capita) reduce profitability of frontier agriculture and slows deforestation.

Historical Causes

Prehistory

The Carboniferous Rainforest Collapse was an event that occurred 300 million years ago. Climate change devastated tropical rainforests causing the extinction of many plant and animal species. The change was abrupt, specifically, at this time climate became cooler and drier, conditions that are not favourable to the growth of rainforests and much of the biodiversity within them. Rainforests were fragmented forming shrinking 'islands' further and further apart. This sudden collapse affected several large groups, effects on amphibians were particularly devastating, while reptiles fared better, being ecologically adapted to the drier conditions that followed.

An array of Neolithic artifacts, including bracelets, axe heads, chisels, and polishing tools.

Rainforests once covered 14% of the earth's land surface; now they cover a mere 6% and experts estimate that the last remaining rainforests could be consumed in less than 40 years. Small scale deforestation was practiced by some societies for tens of thousands of years before the beginnings of civilization. The first evidence of deforestation appears in the Mesolithic period. It was probably used to convert closed forests into more open ecosystems favourable to game animals. With the advent of agriculture, larger areas began to be deforested, and fire became the prime tool to clear land for crops. In Europe there is little solid evidence before 7000 BC. Mesolithic foragers used fire to create openings for red deer and wild boar. In Great Britain, shade-tolerant species such as oak and ash are replaced in the pollen record by hazels, brambles, grasses and nettles. Removal of the forests led to decreased transpiration, resulting in the formation of upland peat bogs. Widespread decrease in elm pollen across Europe between 8400–8300 BC and 7200–7000 BC, starting in southern Europe and gradually moving north to Great Britain, may represent land clearing by fire at the onset of Neolithic agriculture.

The Neolithic period saw extensive deforestation for farming land. Stone axes were being made from about 3000 BC not just from flint, but from a wide variety of hard rocks from across Britain and North America as well. They include the noted Langdale axe industry in the English Lake District, quarries developed at Penmaenmawr in North Wales and numerous other locations. Rough-outs were made locally near the quarries, and some were polished locally to give a fine finish. This step not only increased the mechanical strength of the axe, but also made penetration of wood easier. Flint was still used from sources such as Grimes Graves but from many other mines across Europe.

Evidence of deforestation has been found in Minoan Crete; for example the environs of the Palace of Knossos were severely deforested in the Bronze Age.

Pre-industrial History

Throughout most of history, humans were hunter gatherers who hunted within forests. In most areas, such as the Amazon, the tropics, Central America, and the Caribbean, only after shortages of wood and other forest products occur are policies implemented to ensure forest resources are used in a sustainable manner.

In ancient Greece, Tjeered van Andel and co-writers summarized three regional studies of historic erosion and alluviation and found that, wherever adequate evidence exists, a major phase of erosion follows, by about 500-1,000 years the introduction of farming in the various regions of Greece, ranging from the later Neolithic to the Early Bronze Age. The thousand years following the mid-first millennium BC saw serious, intermittent pulses of soil erosion in numerous places. The historic silting of ports along the southern coasts of Asia Minor (*e.g.* Clarus, and the examples of Ephesus, Priene and Miletus, where harbors had to be abandoned because of the silt deposited by the Meander) and in coastal Syria during the last centuries BC.

Easter Island

Easter Island has suffered from heavy soil erosion in recent centuries, aggravated by agriculture and deforestation. Jared Diamond gives an extensive look into the collapse of the ancient Easter Islanders in his book *Collapse*. The disappearance of the island's trees seems to coincide with a decline of its civilization around the 17th and 18th century. He attributed the collapse to deforestation and over-exploitation of all resources.

The famous silting up of the harbor for Bruges, which moved port commerce to Antwerp, also followed a period of increased settlement growth (and apparently of deforestation) in the upper river basins. In early medieval Riez in upper Provence, alluvial silt from two small rivers raised the riverbeds and widened the floodplain, which slowly buried the Roman settlement in alluvium and gradually moved new construction to higher ground; concurrently the headwater valleys above Riez were being opened to pasturage.

A typical progress trap was that cities were often built in a forested area, which would provide wood for some industry (for example, construction, shipbuilding, pottery). When deforestation occurs without proper replanting, however; local wood supplies become difficult to obtain near enough to remain competitive, leading to the city's abandonment, as happened repeatedly in Ancient Asia Minor. Because of fuel needs, mining and metallurgy often led to deforestation and city abandonment.

With most of the population remaining active in (or indirectly dependent on) the agricultural sector, the main pressure in most areas remained land clearing for crop and cattle farming. Enough wild green was usually left standing (and partially used, for example, to collect firewood, timber and fruits, or to graze pigs) for wildlife to remain viable. The elite's (nobility and higher clergy) protection of their own hunting privileges and game often protected significant woodlands.

Major parts in the spread (and thus more durable growth) of the population were played by monastical 'pioneering' (especially by the Benedictine and Commercial orders) and some feudal lords' recruiting farmers to settle (and become tax payers) by offering relatively good legal and fiscal conditions. Even when speculators sought to encourage towns, settlers needed an agricultural belt around or sometimes within defensive walls. When populations were quickly decreased by causes such as the Black Death or devastating warfare (for example, Genghis Khan's Mongol hordes in eastern and central Europe, Thirty Years' War in Germany), this could lead to settlements being abandoned. The land was reclaimed by nature, but the secondary forests usually lacked the original biodiversity.

Deforestation of Brazil's Atlantic Forest c.1820-1825.

From 1100 to 1500 AD, significant deforestation took place in Western Europe as a result of the expanding human population. The large-scale building of wooden sailing ships by European (coastal) naval owners since the 15th century for exploration, colonisation, slave trade–and other trade on the high seas consumed many forest resources. Piracy also contributed to the over harvesting of forests, as in Spain. This led to a weakening of the domestic economy after Columbus' discovery of America, as the economy became dependent on colonial activities (plundering, mining, cattle, plantations, trade, etc.)

In *Changes in the Land* (1983), William Cronon analyzed and documented 17th-century English colonists' reports of increased seasonal flooding in New England during the period when new settlers initially cleared the forests for agriculture. They believed flooding was linked to widespread forest clearing upstream.

The massive use of charcoal on an industrial scale in Early Modern Europe was a new type of consumption of western forests; even in Stuart England, the relatively primitive production of charcoal has already reached an impressive level. Stuart England was so widely deforested that it depended on the Baltic trade for ship timbers, and looked to the untapped forests of New England to supply the need. Each of Nelson's Royal Navy war ships at Trafalgar (1805) required 6,000 mature oaks for its construction. In France, Colbert planted oak forests to supply the French navy in the future. When the oak plantations matured in the mid-19th century, the masts were no longer required because shipping had changed.

Norman F. Cantor's summary of the effects of late medieval deforestation applies equally well to Early Modern Europe:

Europeans had lived in the midst of vast forests throughout the earlier medieval centuries. After 1250 they became so skilled at deforestation that by 1500 they were running short of wood for heating and cooking. They were faced with a nutritional decline because of the elimination of the generous supply of wild game that had inhabited the now-disappearing forests, which throughout medieval times had provided the staple of their carnivorous high-protein diet. By 1500 Europe was on the edge of a fuel and nutritional disaster [from] which it was saved in the sixteenth century only by the burning of soft coal and the cultivation of potatoes and maize.

Industrial Era

In the 19th century, introduction of steamboats in the United States was the cause of deforestation of banks of major rivers, such as the Mississippi River, with increased and more severe flooding one of the environmental results. The steamboat crews cut wood every day from the riverbanks to fuel the steam engines. Between St. Louis and the confluence with the Ohio River to the south, the Mississippi became more wide and shallow, and changed its channel laterally. Attempts to improve navigation by the use of snag pullers often resulted in crews' clearing large trees 100 to 200 feet (61 m) back from the banks. Several French colonial towns of the Illinois Country, such as Kaskaskia, Cahokia and St. Philippe, Illinois were flooded and abandoned in the late 19th century, with a loss to the cultural record of their archeology.

The wholescale clearance of woodland to create agricultural land can be seen in many parts of the world, such as the Central forest-grasslands transition and other areas of the Great Plains of the United States. Specific parallels are seen in the 20th-century deforestation occurring in many developing nations.

Rates of Deforestation

Global deforestation sharply accelerated around 1852. It has been estimated that about half of the Earth's mature tropical forests—between 7.5 million and 8 million km^2 (2.9 million to 3 million sq mi) of the original 15 million to 16 million km^2 (5.8 million to 6.2 million sq mi) that until 1947 covered the planet—have now been destroyed. Some scientists have predicted that unless significant measures (such as seeking out and protecting old growth forests that have not been disturbed) are taken on a worldwide basis, by 2030 there will only be 10% remaining, with another 10% in a degraded condition. 80% will have been lost, and with them hundreds of thousands of irreplaceable species. Some cartographers have attempted to illustrate the sheer scale of deforestation by country using a cartogram.

Slash-and-burn farming in the state of Rondônia, western Brazil.

Estimates vary widely as to the extent of tropical deforestation. Scientists estimate that one fifth of the world's tropical rainforest was destroyed between 1960 and 1990. They claim that that rainforests 60 years ago covered 14% of the world's land surface, now only cover 5–7%, and that all tropical forests will be gone by the middle of the 21st century.

A 2002 analysis of satellite imagery suggested that the rate of deforestation in the humid tropics (approximately 5.8 million hectares per year) was roughly 23% lower than the most commonly quoted rates. Conversely, a newer analysis of satellite images reveals that deforestation of the Amazon rainforest is twice as fast as scientists previously estimated.

Some have argued that deforestation trends may follow a Kuznets curve, which if true would nonetheless fail to eliminate the risk of irreversible loss of non-economic forest values (for example, the extinction of species).

Satellite image of Haiti's border with the Dominican Republic shows the amount of deforestation on the Haitian side.

A 2005 report by the United Nations Food and Agriculture Organization (FAO) estimates that although the Earth's total forest area continues to decrease at about 13 million hectares per year, the global rate of deforestation has recently been slowing. Still others claim that rainforests are being destroyed at an ever-quickening pace. The London-based Rainforest Foundation notes that "the UN figure is based on a definition of forest as being an area with as little as 10% actual tree cover, which would therefore include areas that are actually savannah-like ecosystems and badly damaged forests." Other critics of the FAO data point out that they do not distinguish between forest types, and that they are based largely on reporting from forestry departments of individual countries, which do not take into account unofficial activities like illegal logging.

Despite these uncertainties, there is agreement that destruction of rainforests remains a significant environmental problem. Up to 90% of West Africa's coastal rainforests have disappeared since 1900. In South Asia, about 88% of the rainforests have been lost. Much of what remains of the world's rainforests is in the Amazon basin, where the Amazon Rainforest covers approximately 4 million square kilometres. The regions with the highest tropical deforestation rate between 2000 and 2005 were Central America—which lost 1.3% of its forests each year—and tropical Asia. In Central America, two-thirds of lowland tropical forests have been turned into pasture since 1950 and 40% of all the rainforests have been lost in the last 40 years. Brazil has lost 90–95% of its Mata Atlântica forest. Paraguay was losing its natural semi humid forests in the country's western regions at a rate of 15.000 hectares at a randomly studied 2-month period in 2010, Paraguay's parliament refused in 2009 to pass a law that would have stopped cutting of natural forests altogether.

Deforestation around Pakke Tiger Reserve, India.

Madagascar has lost 90% of its eastern rainforests. As of 2007, less than 1% of Haiti's forests remained. Mexico, India, the Philippines, Indonesia, Thailand, Burma, Malaysia, Bangladesh, China, Sri Lanka, Laos, Nigeria, the Democratic Republic of the Congo, Liberia, Guinea, Ghana and the Ivory Coast, have lost large areas of their rainforest. Several countries, notably Brazil, have declared their deforestation a national emergency. The World Wildlife Fund's ecoregion project catalogues habitat types throughout the world, including habitat loss such as deforestation, showing for example that even in the rich forests of parts of Canada such as the Mid-Continental Canadian forests of the prairie provinces half of the forest cover has been lost or altered.

Regions

Rates of deforestation vary around the world.

In 2011 Conservation International listed the top 10 most endangered forests, characterized by having all lost 90% or more of their original habitat, and each harboring at least 1500 endemic plant species (species found nowhere else in the world).

Top 10 Most Endangered Forests 2011				
Endangered forest	**Region**	**Remaining habitat**	**Predominate vegetation type**	**Notes**
Indo-Burma	Asia-Pacific	5%	Tropical and subtropical moist broadleaf forests	Rivers, floodplain wetlands, mangrove forests. Burma, Thailand, Laos, Vietnam, Cambodia, India.
New Caledonia	Asia-Pacific	5%	Tropical and subtropical moist broadleaf forests	
Sundaland	Asia-Pacific	7%	Tropical and subtropical moist broadleaf forests	Western half of the Indo-Malayan archipelago including southern Borneo and Sumatra.
Philippines	Asia-Pacific	7%	Tropical and subtropical moist broadleaf forests	Forests over the entire country including 7,100 islands.
Atlantic Forest	South America	8%	Tropical and subtropical moist broadleaf forests	Forests along Brazil's Atlantic coast, extends to parts of Paraguay, Argentina and Uruguay.
Mountains of Southwest China	Asia-Pacific	8%	Temperate coniferous forest	
California Floristic Province	North America	10%	Tropical and subtropical dry broadleaf forests	
Coastal Forests of Eastern Africa	Africa	10%	Tropical and subtropical moist broadleaf forests	Mozambique, Tanzania, Kenya, Somalia.
Madagascar & Indian Ocean Islands	Africa	10%	Tropical and subtropical moist broadleaf forests	Madagascar, Mauritius, Reunion, Seychelles, Comoros.
Eastern Afromontane	Africa	11%	Tropical and subtropical moist broadleaf forests Montane grasslands and shrublands	Forests scattered along the eastern edge of Africa, from Saudi Arabia in the north to Zimbabwe in the south.

Control

Reducing Emissions

Main international organizations including the United Nations and the World Bank, have begun to develop programs aimed at curbing deforestation. The blanket term Reducing Emissions from Deforestation and Forest Degradation (REDD) describes these sorts of programs, which use direct monetary or other incentives to encourage developing countries to limit and/or roll back deforestation. Funding has been an issue, but at the UN Framework Convention on Climate Change (UNFCCC) Conference of the Parties-15 (COP-15) in Copenhagen in December 2009, an accord was reached with a collective commitment by developed countries for new and additional resources, including forestry and investments through international institutions, that will approach USD 30 billion for the period 2010–2012. Significant work is underway on tools for use in monitoring developing country adherence to their agreed REDD targets. These tools, which rely on remote forest monitoring using satellite imagery and other data sources, include the Center for Global Development's FORMA (Forest Monitoring for Action) initiative and the Group on Earth Observations' Forest Carbon Tracking Portal. Methodological guidance for forest monitoring was also emphasized at COP-15. The environmental organization Avoided Deforestation Partners leads the campaign for development of REDD through funding from the U.S. government. In 2014, the Food and Agriculture Organization of the United Nations and partners launched Open Foris - a set of open-source software tools that assist countries in gathering, producing and disseminating information on the state of forest resources. The tools support the inventory lifecycle, from needs assessment, design, planning, field data collection and management, estimation analysis, and dissemination. Remote sensing image processing tools are included, as well as tools for international reporting for Reducing emissions from deforestation and forest degradation (REDD) and MRV and FAO's Global Forest Resource Assessments.

In evaluating implications of overall emissions reductions, countries of greatest concern are those categorized as High Forest Cover with High Rates of Deforestation (HFHD) and Low Forest Cover with High Rates of Deforestation (LFHD). Afghanistan, Benin, Botswana, Burma, Burundi, Cameroon, Chad, Ecuador, El Salvador, Ethiopia, Ghana, Guatemala, Guinea, Haiti, Honduras, Indonesia, Liberia, Malawi, Mali, Mauritania, Mongolia, Namibia, Nepal, Nicaragua, Niger, Nigeria, Pakistan, Paraguay, Philippines, Senegal, Sierra Leone, Sri Lanka, Sudan, Togo, Uganda, United Republic of Tanzania, Zimbabwe are listed as having Low Forest Cover with High Rates of Deforestation (LFHD). Brazil, Cambodia, Democratic Peoples Republic of Korea, Equatorial Guinea, Malaysia, Solomon Islands, Timor-Leste, Venezuela, Zambia are listed as High Forest Cover with High Rates of Deforestation (HFHD).

Payments for Conserving Forests

In Bolivia, deforestation in upper river basins has caused environmental problems, including soil erosion and declining water quality. An innovative project to try and remedy this situation involves landholders in upstream areas being paid by downstream water users to conserve forests. The landholders receive US$20 to conserve the trees, avoid polluting livestock practices, and enhance the biodiversity and forest carbon on their land. They also receive US$30, which purchases a beehive, to compensate for conservation for two hectares of water-sustaining forest for five years. Honey revenue per hectare of forest is US$5 per year, so within five years, the landholder has sold US$50 of honey. The project is being conducted by Fundación Natura Bolivia and Rare Conservation, with support from the Climate & Development Knowledge Network.

Farming

New methods are being developed to farm more intensively, such as high-yield hybrid crops, greenhouse, autonomous building gardens, and hydroponics. These methods are often dependent on chemical inputs to maintain necessary yields. In cyclic agriculture, cattle are grazed on farm land that is resting and rejuvenating. Cyclic agriculture actually increases the fertility of the soil. Intensive farming can also decrease soil nutrients by consuming at an accelerated rate the trace minerals needed for crop growth.The most promising approach, however, is the concept of food forests in permaculture, which consists of agroforestal systems carefully designed to mimic natural forests, with an emphasis on plant and animal species of interest for food, timber and other uses. These systems have low dependence on fossil fuels and agro-chemicals, are highly self-maintaining, highly productive, and with strong positive impact on soil and water quality, and biodiversity.

Monitoring Deforestation

There are multiple methods that are appropriate and reliable for reducing and monitoring deforestation. One method is the "visual interpretation of aerial photos or satellite imagery that is labor-intensive but does not require high-level training in computer image processing or extensive computational resources". Another method includes hot-spot analysis (that is, locations of rapid change) using expert opinion or coarse resolution satellite data to identify locations for detailed digital analysis with high resolution satellite images. Deforestation is typically assessed by quantifying the amount of area deforested, measured at the present time. From an environmental point of view, quantifying the damage and its possible consequences is a more important task, while conservation efforts are more focused on forested land protection and development of land-use alternatives to avoid continued deforestation. Deforestation rate and total area deforested, have been widely used for monitoring deforestation in many regions, including the Brazilian Amazon deforestation monitoring by INPE.

Forest Management

Efforts to stop or slow deforestation have been attempted for many centuries because it has long been known that deforestation can cause environmental damage sufficient in some cases to cause societies to collapse. In Tonga, paramount rulers developed policies designed to prevent conflicts between short-term gains from converting forest to farmland and long-term problems forest loss would cause, while during the 17th and 18th centuries in Tokugawa, Japan, the shoguns developed a highly sophisticated system of long-term planning to stop and even reverse deforestation of the preceding centuries through substituting timber by other products and more efficient use of land that had been farmed for many centuries. In 16th-century Germany, landowners also developed silviculture to deal with the problem of deforestation. However, these policies tend to be limited to environments with *good rainfall, no dry season* and *very young soils* (through volcanism or glaciation). This is because on older and less fertile soils trees grow too slowly for silviculture to be economic, whilst in areas with a strong dry season there is always a risk of forest fires destroying a tree crop before it matures.

In the areas where "slash-and-burn" is practiced, switching to "slash-and-char" would prevent the rapid deforestation and subsequent degradation of soils. The biochar thus created, given back to the soil, is not only a durable carbon sequestration method, but it also is an extremely beneficial amendment to the soil. Mixed with biomass it brings the creation of terra preta, one of the richest soils on the planet and the only one known to regenerate itself.

Sustainable Practices

Certification, as provided by global certification systems such as Programme for the Endorsement of Forest Certification and Forest Stewardship Council, contributes to tackling deforestation by creating market demand for timber from sustainably managed forests. According to the United Nations Food and Agriculture Organization (FAO), "A major condition for the adoption of sustainable forest management is a demand for products that are produced sustainably and consumer willingness to pay for the higher costs entailed. Certification represents a shift from regulatory approaches to market incentives to promote sustainable forest management. By promoting the positive attributes of forest products from sustainably managed forests, certification focuses on the demand side of environmental conservation." Rainforest Rescue argues that the standards of organizations like FSC are too closely connected to timber industry interests and therefore do not guarantee environmentally and socially responsible forest management. In reality, monitoring systems are inadequate and various cases of fraud have been documented worldwide.

Bamboo is advocated as a more sustainable alternative for cutting down wood for fuel.

Some nations have taken steps to help increase the amount of trees on Earth. In 1981, China created National Tree Planting Day Forest and forest coverage had now reached 16.55% of China's land mass, as against only 12% two decades ago.

Using fuel from bamboo rather than wood results in cleaner burning, and since bamboo matures much faster than wood, deforestation is reduced as supply can be replenished faster.

Reforestation

In many parts of the world, especially in East Asian countries, reforestation and afforestation are increasing the area of forested lands. The amount of woodland has increased in 22 of the world's 50 most forested nations. Asia as a whole gained 1 million hectares of forest between 2000 and 2005. Tropical forest in El Salvador expanded more than 20% between 1992 and 2001. Based on these trends, one study projects that global forest will increase by 10%—an area the size of India—by 2050.

In the People's Republic of China, where large scale destruction of forests has occurred, the government has in the past required that every able-bodied citizen between the ages of 11 and 60 plant three to five trees per year or do the equivalent amount of work in other forest services. The government claims that at least 1 billion trees have been planted in China every year since 1982. This is no longer required today, but March 12 of every year in China is the Planting Holiday. Also, it has introduced the Green Wall of China project, which aims to halt the expansion of the Gobi desert through the planting of trees. However, due to the large percentage of trees dying off after planting (up to 75%), the project is not very successful. There has been a 47-million-hectare increase in forest area in China since the 1970s. The total number of trees amounted to be about 35 billion and 4.55% of China's land mass increased in forest coverage. The forest coverage was 12% two decades ago and now is 16.55%.

An ambitious proposal for China is the Aerially Delivered Re-forestation and Erosion Control System and the proposed Sahara Forest Project coupled with the Seawater Greenhouse.

In Western countries, increasing consumer demand for wood products that have been produced and harvested in a sustainable manner is causing forest landowners and forest industries to become increasingly accountable for their forest management and timber harvesting practices.

The Arbor Day Foundation's Rain Forest Rescue program is a charity that helps to prevent deforestation. The charity uses donated money to buy up and preserve rainforest land before the lumber companies can buy it. The Arbor Day Foundation then protects the land from deforestation. This also locks in the way of life of the primitive tribes living on the forest land. Organizations such as Community Forestry International, Cool Earth, The Nature Conservancy, World Wide Fund for Nature, Conservation International, African Conservation Foundation and Greenpeace also focus on preserving forest habitats. Greenpeace in particular has also mapped out the forests that are still intact and published this information on the internet. World Resources Institute in turn has made a simpler thematic map showing the amount of forests present just before the age of man (8000 years ago) and the current (reduced) levels of forest. These maps mark the amount of afforestation required to repair the damage caused by people.

Forest Plantations

To meet the world's demand for wood, it has been suggested by forestry writers Botkins and Sedjo that high-yielding forest plantations are suitable. It has been calculated that plantations yielding 10 cubic meters per hectare annually could supply all the timber required for international trade on 5% of the world's existing forestland. By contrast, natural forests produce about 1–2 cubic meters per hectare; therefore, 5–10 times more forestland would be required to meet demand. Forester Chad Oliver has suggested a forest mosaic with high-yield forest lands interspersed with conservation land.

In the country of Senegal, on the western coast of Africa, a movement headed by youths has helped to plant over 6 million mangrove trees. The trees will protect local villages from storm damages and will provide a habitat for local wildlife. The project started in 2008, and already the Senegalese government has been asked to establish rules and regulations that would protect the new mangrove forests.

Military Context

American Sherman tanks knocked out by Japanese artillery on Okinawa.

While the preponderance of deforestation is due to demands for agricultural and urban use for the human population, there are some examples of military causes. One example of deliberate deforestation is that which took place in the U.S. zone of occupation in Germany after World War II. Before the onset of the Cold War, defeated Germany was still considered a potential future threat rather than potential future ally. To address this threat, attempts were made to lower German industrial potential, of which forests were deemed an element. Sources in the U.S. government admitted that the purpose of this was that the "ultimate destruction of the war potential of German forests." As a consequence of the practice of clear-felling, deforestation resulted which could "be replaced only by long forestry development over perhaps a century."

Deforestation can also be one consequence of war. For example, in the 1945 Battle of Okinawa, bombardment and other combat operations reduced the lush tropical landscape into "a vast field of mud, lead, decay and maggots". Deforestation can also be an intentional tactic of military forces. Defoliants (like Agent Orange or others) was used by the British in the Malayan Emergency, and by the United States in the Korean War and Vietnam War.

Public Health Context

Deforestation eliminates a great number of species of plants and animals which also often results in an increase in disease. Loss of native species allows new species to come to dominance. Often the destruction of predatory species can result in an increase in rodent populations. These are known to carry plagues. Additionally, erosion can produce pools of stagnant water that are perfect breeding grounds for mosquitos, well known vectors of malaria, yellow fever, nipah virus, and more. Deforestation can also create a path for non-native species to flourish such as certain types of snails, which have been correlated with an increase in schistosomiasis cases.

Deforestation is occurring all over the world and has been coupled with an increase in the occurrence of disease outbreaks. In Malaysia, thousands of acres of forest have been cleared for pig farms. This has resulted in an increase in the zoonosis the Nipah virus. In Kenya, deforestation has led to an increase in malaria cases which is now the leading cause of morbidity and mortality the country.

Another pathway through which deforestation affects disease is the relocation and dispersion of disease-carrying hosts. This disease emergence pathway can be called "range expansion," whereby the host's range (and thereby the range of pathogens) expands to new geographic areas. Through deforestation, hosts and reservoir species are forced into neighboring habitats. Accompanying the reservoir species are pathogens that have the ability to find new hosts in previously unexposed regions. As these pathogens and species come into closer contact with humans, they are infected both directly and indirectly.

A catastrophic example of range expansion is the 1998 outbreak of Nipah Virus in Malaysia. For a number of years, deforestation, drought, and subsequent fires led to a dramatic geographic shift and density of fruit bats, a reservoir for Nipah virus. Deforestation reduced the available fruiting trees in the bats' habitat, and they encroached on surrounding orchards which also happened to be the location of a large number of pigsties. The bats, through proximity spread the Nipah to pigs. While the virus infected the pigs, mortality was much lower than among humans, making the pigs a virulent host leading to the transmission of the virus to humans. This resulted in 265 reported cases of encephalitis, of which 105 resulted in death. This example provides an important lesson for the impact deforestation can have on human health.

Another example of range expansion due to deforestation and other anthropogenic habitat impacts includes the Capybara rodent in Paraguay. This rodent is the host of a number of zoonotic diseases and, while there has not yet been a human-borne outbreak due to the movement of this rodent into new regions, it offers an example of how habitat destruction through deforestation and subsequent movements of species is occurring regularly.

A now well-developed theory is that the spread of HIV it is at least partially due deforestation. Rising populations created a food demand and with deforestation opening up new areas of the forest the hunters harvested a great deal of primate bushmeat, which is believed to be the origin of HIV.

Indirect Land Use Change Impacts of Biofuels

The indirect land use change impacts of biofuels, also known as ILUC, relates to the unintended consequence of releasing more carbon emissions due to land-use changes around the world induced by the expansion of croplands for ethanol or biodiesel production in response to the increased global demand for biofuels.

Brazilian cerrado.

As farmers worldwide respond to higher crop prices in order to maintain the global food supply-and-demand balance, pristine lands are cleared to replace the food crops that were diverted elsewhere to biofuels' production. Because natural lands, such as rainforests and grasslands, store carbon in their soil and biomass as plants grow each year, clearance of wilderness for new farms translates to a net increase in greenhouse gas emissions. Due to this change in the carbon stock of the soil and the biomass, indirect land use change has consequences in the GHG balance of a biofuel.

Other authors have also argued that indirect land use changes produce other significant social and environmental impacts, affecting biodiversity, water quality, food prices and supply, land tenure, worker migration, and community and cultural stability.

Amazon rainforest.

History

The estimates of carbon intensity for a given biofuel depend on the assumptions regarding several variables. As of 2008, multiple full life cycle studies had found that corn ethanol, cellulosic ethanol and Brazilian sugarcane ethanol produce lower greenhouse gas emissions than gasoline. None of these studies, however, considered the effects of indirect land-use changes, and though land use impacts were acknowledged, estimation was considered too complex and difficult to model. A controversial paper published in February 2008 in Sciencexpress by a team led by Searchinger from Princeton University concluded that such effects offset the (positive) direct effects of both corn and cellulosic ethanol and that Brazilian sugarcane performed better, but still resulted in a small carbon debt.

Malaysian cloud forest.

After the Searchinger team paper, estimation of carbon emissions from ILUC, together with the food vs. fuel debate, became one of the most contentious issues relating to biofuels, debated in the popular media, scientific journals, op-eds and public letters from the scientific community, and the ethanol industry, both American and Brazilian. This controversy intensified in April 2009 when the California Air Resources Board (CARB) set rules that included ILUC impacts to establish the California Low-Carbon Fuel Standard that entered into force in 2011.

In May 2009 U.S. Environmental Protection Agency (EPA) released a notice of proposed rulemaking for implementation of the 2007 modification of the Renewable Fuel Standard (RFS). EPA's proposed regulations also included ILUC, causing additional controversy among ethanol producers. EPA's February 3, 2010 final rule incorporated ILUC based on modelling that was significantly improved over the initial estimates.

The UK Renewable Transport Fuel Obligation program requires the Renewable Fuels Agency (RFA) to report potential indirect impacts of biofuel production, including indirect land use change or changes to food and other commodity prices. A July 2008 RFA study, known as the Gallager Review, found several risks and uncertainties, and that the "quantification of GHG emissions from indirect land-use change requires subjective assumptions and contains considerable uncertainty", and required further examination to properly incorporate indirect effects into calculation methodologies. A similarly cautious approach was followed by the European Union. In December 2008 the European Parliament adopted more stringent sustainability criteria for biofuels and directed the European Commission to develop a methodology to factor in GHG emissions from indirect land use change.

Studies and Controversy

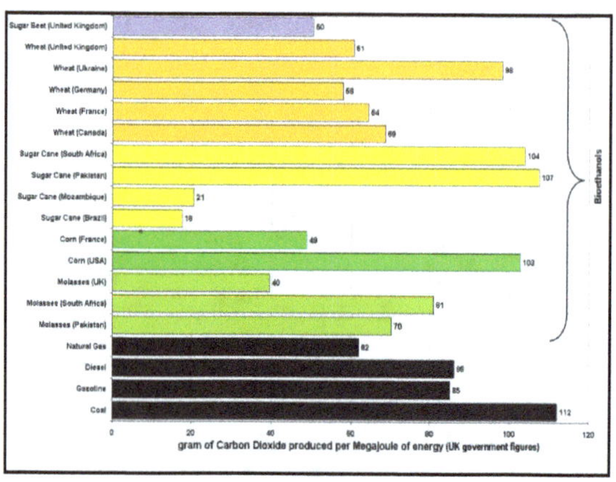

UK figures for the carbon intensity of bioethanol and fossil fuels. Graph assumes all bioethanols are burnt in country of origin and prior cropland was used to grow feedstock. No ILUC effects were included.

Before 2008, several full life cycle ("Well to Wheels" or WTW) studies had found that corn ethanol reduced transport-related greenhouse gas emissions. In 2007 a University of California, Berkeley team led by Farrel evaluated six previous studies, concluding that corn ethanol reduced GHG emissions by only 13 percent. However, 20 to 30 percent reduction for corn ethanol, and 85 to 85 percent for cellulosic ethanol, both figures estimated by Wang from Argonne National Laboratory, are more commonly cited. Wang reviewed 22 studies conducted between

1979 and 2005, and ran simulations with Argonne's GREET model. These studies accounted for direct land use changes. Several studies of Brazilian sugarcane ethanol showed that sugarcane as feedstock reduces GHG by 86 to 90 percent given no significant land use change. Estimates of carbon intensity depend on crop productivity, agricultural practices, power sources for ethanol distilleries and the energy efficiency of the distillery. None of these studies considered ILUC, due to estimation difficulties. Preliminary estimates by Delucchi from the University of California, Davis, suggested that carbon released by new lands converted to agricultural use was a large percentage of life-cycle emissions.

Searchinger and Fargione Studies

In 2008 Timothy Searchinger, a lawyer from Environmental Defense Fund, concluded that ILUC affects the life cycle assessment and that instead of saving, both corn and cellulosic ethanol increased carbon emissions as compared to gasoline by 93 and 50 percent respectively. Ethanol from Brazilian sugarcane performed better, recovering initial carbon emissions in 4 years, while U.S. corn ethanol required 167 years and cellulosic ethanol required a 52 years payback period. The study limited the analysis a 30-year period, assuming that land conversion emits 25 percent of the carbon stored in soils and all carbon in plants cleared for cultivation. Brazil, China, and India were considered among the overseas locations where land use change would occur as a result of diverting U.S. corn cropland, and it was assumed that new cropland in each of these regions correspond to different types of forest, savanna or grassland based on the historical proportion of each converted to cultivation in these countries during the 1990s.

Summary of Searchinger et al. comparison of corn ethanol and gasoline GHG emissions with and without land use change (Grams of CO_2 released per megajoule of energy in fuel)

Fuel type (U.S.)	Carbon intensity	Reduction GHG	Carbon intensity + ILUC	Reduction GHG
Gasoline	**92**	-	**92**	-
Corn ethanol	74	-20%	177	+93%
Cellulosic ethanol	28	-70%	138	+50%

Notes: Calculated using default assumptions for 2015 scenario for ethanol in E85. Gasoline is a combination of conventional and reformulated gasoline.

Fargione and his team published a separate paper in the same issue of Sciencexpress claiming that clearing lands to produce biofuel feedstock created a carbon deficit. This deficit applies to both direct and indirect land use changes. The study examined six conversion scenarios: Brazilian Amazon to soybean biodiesel, Brazilian Cerrado to soybean biodiesel, Brazilian Cerrado to sugarcane ethanol, Indonesian or Malaysian lowland tropical rainforest to palm biodiesel, Indonesian or Malaysian peatland tropical rainforest to palm biodiesel, and U.S. Central grassland to corn ethanol. The carbon debt was defined as the amount of CO_2 released during the first 50 years of this process of land conversion. For the two most common ethanol feedstocks, the study found that sugarcane ethanol produced on natural cerrado lands would take about 17 years to repay its carbon debt, while corn ethanol produced on U.S. central grasslands would result in a repayment time of about 93 years. The worst-case scenario is converting Indonesian or Malaysian tropical peatland rainforest to palm biodiesel production, which would require about 420 years to repay.

Criticism and Controversy

The Searchinger and Fargione studies created controversy in both the popular media and in scientific journals. Robert Zubrin observed that Searchinger's "indirect analysis" approach is pseudo-scientific and can be used to "prove anything".

Wang and Haq from Argonne National Laboratory claiming: the assumptions were outdated; they ignored the potential of increased efficiency; and no evidence showed that "U.S. corn ethanol production has so far caused indirect land use in other countries." They concluded that Searchinger demonstrated that ILUC "is much more difficult to model than direct land use changes". In his response Searchinger rebutted each technical objection and asserted that "... any calculation that ignores these emissions, however challenging it is to predict them with certainty, is too incomplete to provide a basis for policy decisions."

Another criticism, by Kline and Dale from Oak Ridge National Laboratory, held that Searchinger et al. and Fargione et al. "... do not provide adequate support for their claim that biofuels cause high emissions due to land-use change", as their conclusions depends on a misleading assumption because more comprehensive field research found that these land use changes "... are driven by interactions among cultural, technological, biophysical, economic, and demographic forces within a spatial and temporal context rather than by a single crop market". Fargione et al. responded in part that although many factors contributed to land clearing, this "observation does not diminish the fact that biofuels also contribute to land clearing if they are produced on existing cropland or on newly cleared lands". Searching disagreed with all of Kline and Dale arguments.

The U.S. biofuel industry also reacted, claiming that the "Searchinger study is clearly a 'worst case scenario' analysis ..." and that this study "relies on a long series of highly subjective assumptions ..." Searchinger rebutted each claim, concluding that NFA's criticisms were invalid. He noted that even if some of his assumptions are high estimates, the study also made many conservative assumptions.

Slash and burn forest removal in Brazil.

Cattle ranching in Brazil.

Brazil

In February 2010, Lapola estimated that planned expansion of Brazilian sugarcane and soybean biofuel plantations through 2020 would replace rangeland, with small direct land-use impact on carbon emissions. However, the expansion of the rangeland frontier into Amazonian forests, driven by cattle ranching, would indirectly offset the savings. "Sugarcane ethanol and soybean biodiesel each contribute to nearly half of the projected indirect deforestation of 121,970 km^2 by 2020, creating a carbon debt that would take about 250 years to be repaid...."

The research also found that oil palm would cause the least land-use changes and associated carbon debt. The analysis also modeled livestock density increases and found that "a higher increase of 0.13 head per hectare in the average livestock density throughout the country could avoid the indirect land-use changes caused by biofuels (even with soybean as the biodiesel feedstock), while still fulfilling all food and bioenergy demands." The authors conclude that intensification of cattle ranching and concentration on oil palm are required to achieve effective carbon savings, recommending closer collaboration between the biofuel and cattle-ranching sectors.

The main Brazilian ethanol industry organization (UNICA) commented that such studies missed the continuing intensification of cattle production already underway.

A study by Arima *et al.* published in May 2011 used spatial regression modeling to provide the first statistical assessment of ILUC for the Brazilian Amazon due to soy production. Previously, the indirect impacts of soy crops were only anecdotal or analyzed through demand models at a global scale, while the study took a regional approach. The analysis showed a strong signal linking the expansion of soybean fields in settled agricultural areas at the southern and eastern rims of the Amazon basin to pasture encroachments for cattle production on the forest frontier. The results demonstrate the need to include ILUC in measuring the carbon footprint of soy crops, whether produced for biofuels or other end-uses.

The Arima study is based on 761 municipalities located in the Legal Amazon of Brazil, and found that between 2003 and 2008, soybean areas expanded by 39,100 km^2 in the basin's agricultural areas, mainly in Mato Grosso. The model showed that a 10% (3,910 km^2) reduction of soy in old pasture areas would have led to a reduction in deforestation of up to 40% (26,039 km^2) in heavily forested municipalities of the Brazilian Amazon. The analysis showed that the displacement of cattle production due to agricultural expansion drives land use change in municipalities located hundreds of kilometers away, and that the Amazonian ILUC is not only measurable but its impact is significant.

Implementation

California LCFS

California carbon intensity values for gasoline, diesel and fuels that substitute them (grams of CO_2 equivalent released per MJ of energy produced)			
Fuel type	**Carbon intensity**	**Carbon intensity + land-use changes**	**Intensity change respect to 2011 LCFS**
Midwest corn ethanol	75.10	105.10	+10%
California gasoline	95.86	95.86	+0.2%
CARB LCFS 2011 for gasoline	-	95.61	-
California diesel (ULSD)	94.71	94.71	+0.2%
CARB LCFS 2011 for diesel	-	94.47	-
California ethanol	50.70	80.70	-16%
Brazilian sugarcane ethanol	27.40	73.40	-23%
Biodiesel (B100) Midwest soybeans	26.93	68.93	-27%
Renewable diesel Midwest soybeans	28.80	68.93	-27%
Cellulosic ethanol (farmed trees)	2.40	20.40	-79%
Compressed natural gas (bio-methane)	11.26	11.26	-88%
Note: The complete lifecycle analysis for these fuels and others evaluated are available at CARB's website. Preliminary values of fuels not included in the 2009 LCFS ruling and subject to refining.			

On April 23, 2009, California Air Resources Board (CARB) approved the specific rules and carbon intensity reference values for the California Low-Carbon Fuel Standard (LCFS) that take effect January 1, 2011. CARB's rulemaking included ILUC. For some biofuels, CARB identified land use changes as a significant source of additional GHG emissions. It established one standard for gasoline and alternative fuels, and a second for diesel fuel and its replacements.

Controversy

The public consultation process before the ruling, and the ruling itself were controversial, yielding 229 comments. ILUC was one of the most contentious issues. On June 24, 2008, 27 scientists and researchers submitted a letter saying, "As researchers and scientists in the field of biomass to biofuel conversion, we are convinced that there simply is not enough hard empirical data to base any sound policy regulation in regards to the indirect impacts of renewable biofuels production. The field is relative new, especially when compared to the vast knowledge base present in fossil fuel production, and the limited analyses are driven by assumptions that sometimes lack robust empirical validation." The New Fuels Alliance, representing more than two-dozen biofuel companies, researchers and investors, questioned the Board intention to include indirect land use change effects into account, wrote "While it is likely true that zero is not the right number for the indirect effects of any product in the real world, enforcing indirect effects in a piecemeal way could have very serious consequences for the LCFS.... The argument that zero is not the right number does not justify enforcing a different wrong number, or penalizing one fuel for one category of indirect effects while giving another fuel pathway a free pass."

On the other side, more than 170 scientists and economists urged that CARB, "include indirect land use change in the lifecycle analyses of heat-trapping emissions from biofuels and other transportation fuels. This policy will encourage development of sustainable, low-carbon fuels that avoid conflict with food and minimize harmful environmental impacts.... There are uncertainties inherent in estimating the magnitude of indirect land use emissions from biofuels, but assigning a value of zero is clearly not supported by the science."

Industry representatives complained that the final rule overstated the environmental effects of corn ethanol, and also criticized the inclusion of ILUC as an unfair penalty to domestic corn ethanol because deforestation in the developing world was being tied to U.S. ethanol production. The 2011 limit for LCFS means that Mid-west corn ethanol failed, unless current carbon intensity was reduced. Oil industry representatives complained that the standard left oil refiners with few options, such as Brazilian sugarcane ethanol, with its accompanying tariff. CARB officials and environmentalists counter that time and economic incentives will allow produces to adapt.

UNICA welcomed the ruling, while urging CARB to better reflect Brazilian practices, lowering their estimates of Brazilian emissions.

The only Board member who voted against the ruling explained that he had a "hard time accepting the fact that we're going to ignore the comments of 125 scientists", referring to the letter submitted by a group of scientists questioning the ILUC penalty. "They said the model was not good enough ... to use at this time as a component part of such an historic new standard." CARB advanced the expected date for an expert working group to report on ILUC with refined estimates from January 2012 to January 2011.

On December 2009 the Renewable Fuels Association (RFA) and Growth Energy, two U.S. ethanol lobbying groups, filed a lawsuit challenging LCFS' constitutionality. The two organizations argued that LCFS violated both the Supremacy Clause and the Commerce Clause, jeopardizing the nationwide ethanol market.

EPA Renewable Fuel Standard

U.S. Environmental Protection Agency Draft life cycle GHG emissions reduction results for different time horizon and discount rate approaches (includes indirect land use change effects)		
Fuel Pathway	**100 years + 2% discount rate**	**30 years + 0% discount rate**
Corn ethanol (natural gas dry mill)(1)	-16%	+5%
Corn ethanol (Best case NG DM)(2)	-39%	-18%

Corn ethanol (coal dry mill)	+13%	+34%
Corn ethanol (biomass dry mill)	-39%	-18%
Corn ethanol (biomass dry mill with combined heat and power)	-47%	-26%
Soybean-based biodiesel	-22%	+4%
Waste grease biodiesel	-80%	-80%
Sugarcane ethanol	-44%	-26%
Cellulosic ethanol from switchgrass	-128%	-124%
Cellulosic ethanol from corn stover	-115%	-116%
Notes: (1) Dry mill (DM) plants grind the entire kernel and generally produce only one primary co-product: distillers grains with solubles (DGS). (2) Best case plants produce wet distillers grains co-product.		

The Energy Independence and Security Act of 2007 (EISA) established new renewable fuel categories and eligibility requirements, setting mandatory lifecycle emissions limits. EISA explicitly mandated EPA to include "direct emissions and significant indirect emissions such as significant emissions from land use changes."

EISA required a 20% reduction in lifecycle GHG emissions for any fuel produced at facilities that commenced construction after December 19, 2007 to be classified as a "renewable fuel"; a 50% reduction for fuels to be classified as "biomass-based diesel" or "advanced biofuel", and a 60% reduction to be classified as "cellulosic biofuel". EISA provided limited flexibility to adjust these thresholds downward by up to 10 percent, and EPA proposed this adjustment for the advanced biofuels category.

On May 5, 2009, EPA released a notice of proposed rulemaking for implementation of the 2007 modification of the Renewable Fuel Standard, known as RFS2. The draft of the regulations was released for public comment during a 60-day period, a public hearing was held on 9 June 2009, and also a workshop was conducted on 10–11 June 2009.

EPA's draft analysis stated that ILUC can produce significant near-term GHG emissions due to land conversion, but that biofuels can pay these back over subsequent years. EPA highlighted two scenarios, varying the time horizon and the discount rate for valuing emissions. The first assumed a 30-year time period uses a 0 percent discount rate (valuing emissions equally regardless of timing). The second scenario used a 100-year time period and a 2% discount rate.

Maize is the main feedstock for the production of ethanol fuel in the U.S..

On the same day that EPA published its notice of proposed rulemaking, President Obama signed a Presidential Directive seeking to advance biofuels research and commercialization. The Directive established the Biofuels Interagency Working Group, to develop policy ideas for increasing investment in next-generation fuels and for reducing their environmental footprint.

Sugarcane is the main feedstock for the production of ethanol fuel in Brazil.

The inclusion of ILUC in the proposed ruling provoked complaints from ethanol and biodiesel producers. Several environmental organizations welcomed the inclusion of ILUC but criticized the consideration of a 100-year payback scenario, arguing that it underestimated land conversion effects. American corn growers, biodiesel producers, ethanol producers and Brazilian sugarcane ethanol producers complained about EPA's methodology, while the oil industry requested an implementation delay.

On June 26, 2009, the House of Representatives approved the American Clean Energy and Security Act 219 to 212, mandating EPA to exclude ILUC for a 5-year period, vis a vis RFS2. During this period, more research is to be conducted to develop more reliable models and methodologies for estimating ILUC, and Congress will review this issue before allowing EPA to rule on this matter. The bill failed in the U.S. Senate.

On February 3, 2010, EPA issued its final RFS2 rule for 2010 and beyond. The rule incorporated direct and significant indirect emissions including ILUC. EPA incorporated comments and data from new studies. Using a 30-year time horizon and a 0% discount rate, EPA concluded that multiple biofuels would meet this standard.

EPA's analysis accepted both ethanol produced from corn starch and biobutanol from corn starch as "renewable fuels". Ethanol produced from sugarcane became an "advanced fuel". Both diesel produced from algal oils and biodiesel from soy oil and diesel from waste oils, fats, and greases fell in the "biomass-based diesel" category. Cellulosic ethanol and cellulosic diesel met the "cellulosic biofuel" standard.

The table summarizes the mean GHG emissions estimated by EPA modelling and the range of variations considering that the main source of uncertainty in the life cycle analysis is the GHG emissions related to international land use change.

U.S. Environmental Protection Agency Life cycle Year 2022 GHG emissions reduction results for RFS2 final rule (includes direct and indirect land use change effects and a 30-year payback period at a 0% discount rate)			
Renewable fuel Pathway (for U.S. consumption)	**Mean GHG emission reduction[1]**	**GHG emission reduction 95% confidence interval[2]**	**Assumptions/comments**
Corn ethanol	21%	7–32%	New or expanded natural gas fired dry mill plant, 37% wet and 63% dry DGS it produces, and employing corn oil fractionation technology.
Corn biobutanol	31%	20–40%	Natural gas fired dry mill plant, 37% wet and 63% dry DGS it produces, and employing corn oil fractionation technology.
Sugarcane ethanol[3]	61%	52–71%	Ethanol is produced and dehydrated in Brazil prior to being imported into the U.S. and the residue is not collected. GHG emissions from ocean tankers hauling ethanol from Brazil to the U.S. are included.
Cellulosic ethanol from switchgrass	110%	102–117%	Ethanol produced using the biochemical process.
Cellulosic ethanol from corn stover	129%	No ILUC	Ethanol produced using the biochemical process. Ethanol produced from agricultural residues does not have any international land use emissions.
Biodiesel from soybean	57%	22–85%	Plant using natural gas.
Waste grease biodiesel	86%	No ILUC	Waste grease feedstock does not have any agricultural or land use emissions.
Notes: (1) Percent reduction in lifecycle GHG emissions compared to the average lifecycle GHG for gasoline or diesel sold or distributed as transportation fuel in 2005. (2) Confidence range accounts for uncertainty in the types of land use change assumptions and the magnitude of resulting GHG emissions. (3) A new Brazil module was develop to model the impact of increased production of Brazilian sugarcane ethanol for use in the U.S. market and the international impacts of Brazilian sugarcane ethanol production. The Brazil module also accounts for the domestic competition between crop and pasture land uses, and allows for livestock intensification (heads of cattle per unit area of land).			

Reactions

UNICA welcomed the ruling, in particular, for the more precise lifecycle emissions estimate and hoped that classification the advanced biofuel designation would help eliminate the tariff.

The U.S. Renewable Fuels Association (RFA) also welcomed the ruling, as ethanol producers "require stable federal policy that provides them the market assurances they need to commercialize new technologies", restating their ILUC objection.

RFA also complained that corn-based ethanol scored only a 21% reduction, noting that without ILUC, corn ethanol achieves a 52% GHG reduction. RFA also objected that Brazilian sugarcane ethanol "benefited disproportionally" because EPA's revisions lowered the initially equal ILUC estimates by half for corn and 93% for sugarcane.

Several Midwestern lawmakers commented that they continued to oppose EPA's consideration of the "dicey science" of indirect land use that "punishes domestic fuels". House Agriculture Chairman Collin Peterson said, "... to think that we can credibly measure the impact of international indirect land use is completely unrealistic, and I will continue to push for legislation that prevents unreliable methods and unfair standards from burdening the biofuels industry."

EPA Administrator Lisa P. Jackson commented that the agency "did not back down from considering land use in its final rules, but the agency took new information into account that led to a more favorable calculation for ethanol". She cited new science and better data on crop yield and productivity, more information on co-products that could be produced from advanced biofuels and expanded land-use data for 160 countries, instead of the 40 considered in the proposed rule.

Europe

As of 2010, European Union and United Kingdom regulators had recognized the need to take ILUC into account, but had not determined the most appropriate methodology.

UK Renewable Transport Fuel Obligation

The UK Renewable Transport Fuel Obligation (RTFO) program requires fuel suppliers to report direct impacts, and asked the Renewable Fuels Agency (RFA) to report potential indirect impacts, including ILUC and commodity price changes. The RFA's July 2008 "Gallager Review", mentioned several risks regarding biofuels and required feedstock production to avoid agricultural land that would otherwise be used for food production, despite concluding that "quantification of GHG emissions from indirect land-use change requires subjective assumptions and contains considerable uncertainty". Some environmental groups argued that emissions from ILUC were not being taken into account and could be creating more emissions.

European Union

On December 17, 2008, the European Parliament approved the Renewable Energy Sources Directive (COM(2008)19) and amendments to the Fuel Quality Directive (Directive 2009/30), which included sustainability criteria for biofuels and mandated consideration of ILUC. The Directive established a 10% biofuel target. A separate Fuel Quality Directive set the EU's Low Carbon Fuel Standard, requiring a 6% reduction in GHG intensity of EU transport fuels by 2020. The legislation ordered the European Commission to develop a methodology to factor in GHG emissions from ILUC by December 31, 2010, based on the best available scientific evidence.

In the meantime, the European Parliament defined lands that were ineligible for producing biofuel feedstocks for the purpose of the Directives. This category included wetlands and continuously forested areas with canopy cover of more than 30 percent or cover between 10 and 30 percent given evidence that its existing carbon stock was low enough to justify conversion.

The Commission subsequently published terms of reference for three ILUC modeling exercises: one using a General Equilibrium model; one using a Partial Equilibrium model and one comparing other global modeling exercises. It also consulted on a limited range of high-level options for addressing ILUC to which 17 countries and 59 organizations responded. The United Nations

Special Rapporteur on the Right to Food and several environmental organizations complained that the 2008 safeguards were inadequate. UNICA called for regulators to establish an empirical and "globally accepted methodology" to consider ILUC, with the participation of researchers and scientists from biofuel crop-producing countries.

In 2010 some NGOs accused the European Commission of lacking transparency given its reluctance to release documents relating to the ILUC work. In March 2010 the Partial and General Equilibrium Modelling results were made available, with the disclaimer that the EC had not adopted the views contained in the materials. These indicate that a 1.25% increase in EU biofuel consumption would require around 5,000,000 hectares (12,000,000 acres) of land globally.

The scenarios for varied from 5.6-8.6% of road transport fuels. The study found that ILUC effects offset part of the emission benefits, and that above the 5.6% threshold, ILUC emissions increase rapidly increase. For the expected scenario of 5.6% by 2020, the study estimated that biodiesel production increases would be mostly domestic, while bioethanol production would take place mainly in Brazil, regardless of EU duties. The analysis concluded that eliminating trade barriers would further reduce emissions, because the EU would import more from Brazil. Under this scenario, "*direct emission savings from biofuels are estimated at 18 Mt CO2, additional emissions from ILUC at 5.3 Mt CO2 (mostly in Brazil), resulting in a global net balance of nearly 13 Mt CO2 savings in a 20 years horizon.* The study also found that ILUC emissions were much greater for biodiesel from vegetable oil and estimated that in 2020 even at the 5.6% level were over half the greenhouse gas emissions from diesel.

As part of the announcement, the Commission stated that it would publish a report on ILUC by the end of 2010.

Certification System

On June 10, 2010, the EC announced its decision to set up certification schemes for biofuels, including imports as part of the Renewable Energy Directive. The Commission encouraged E.U. nations, industry and NGOs to set up voluntary certification schemes. EC figures for 2007 showed that 26% of biodiesel and 31% of bioethanol used in the E.U. was imported, mainly from Brazil and the United States.

Reactions

UNICA welcomed the EU efforts to "engage independent experts in its assessments" but requested that improvements because "... the report currently contains a certain number of inaccuracies, so once these are corrected, we anticipate even higher benefits resulting from the use of Brazilian sugarcane ethanol." UNICA highlighted the fact that the report assumed land expansion that "does not take into consideration the agro-ecological zoning for sugarcane in Brazil, which prevents cane from expanding into any type of native vegetation."

Critics said the 10% figure was reduced to 5.6% of transport fuels partly by exaggerating the contribution of electric vehicles (EV) in 2020, as the study assumed EVs would represent 20% of new car sales, two and six times the car industry's own estimate. They also claimed the study "exaggerates to around 45 percent the contribution of bioethanol—the greenest of all biofuels—and consequently downplays the worst impacts of biodiesel."

Environmental groups found that the measures "are too weak to halt a dramatic increase in deforestation". According to Greenpeace, "indirect land-use change impacts of biofuel production still are not properly addressed", which for them was the most dangerous problem of biofuels.

Industry representatives welcomed the certification system and some dismissed concerns regarding the lack of land use criteria. UNICA and other industry groups wanted the gaps in the rules filled to provide a clear operating framework.

The negotiations between the European Parliament and the Council of European Ministers continue. A deal is not foreseen before 2014.

References

- "Indirect land use change – Possible elements of a policy approach – preparatory draft for stakeholder/expert comments" (PDF). European Commission. Retrieved 05, May 2020
- International Food Policy Research Institute (IFPRI) (2010-03-25). "Global Trade and Environmental Impact Study of the EU Biofuels Mandate" (PDF). Directorate General for Trade of the European Commission. Retrieved 09, February 2020
- Inslee, Jay; Bracken Hendricks (2007). "6. Homegrown Energy". Apollo's Fire. Island Press, Washington, D.C. pp. 153–155, 160–161. ISBN 978-1-59726-175-3
- Taylor, Leslie (2004). The Healing Power of Rainforest Herbs: A Guide to Understanding and Using Herbal Medicinals. Square One. ISBN 9780757001444
- Sperling, Daniel; Deborah Gordon (2009). "Two billion cars: driving toward sustainability". Oxford University Press, New York: 98–99. ISBN 978-0-19-537664-7
- Goettemoeller, Jeffrey; Adrian Goettemoeller (2007). "Sustainable Ethanol: Biofuels, Biorefineries, Cellulosic Biomass, Flex-Fuel Vehicles, and Sustainable Farming for Energy Independence". Prairie Oak Publishing, Maryville, Missouri: 40–41. ISBN 978-0-9786293-0-4
- "Analysis Finds That First-Generation Biofuel Use of Up to 5.6% in EU Road Transport Fuels Delivers Net GHG Emissions Benefits After Factoring in Indirect Land Use Change". Green Car Congress. 2010-03-26. Retrieved 25, July 2020
- Jim Lane (2009-02-24). "CARB votes 9-1 for California Low Carbon Fuel Standard; moves up indirect land use review to Jan 2011 in response to outcry on ILUC". BiofuelsDigest. Retrieved 17, April 2020
- "Renewable Fuel Standard Program(RFS2) Regulatory Impact Analysis" (PDF). U.S. Environmental Protection Agency. February 2010. Archived from the original (PDF) on February 2, 2011
- Ron Nielsen, The Little Green Handbook: Seven Trends Shaping the Future of Our Planet, Picador, New York (2006) ISBN 978-0-312-42581-4
- Christine Moser; Tina Hildebrandt; Robert Bailis (14 November 2013). "International Sustainability Standards and Certification". In Barry D. Solomon. Sustainable Development of Biofuels in Latin America and the Caribbean. Robert Bailis. Springer New York. pp. 27–69. ISBN 978-1-4614-9274-0
- Rudel, T.K. 2005 "Tropical Forests: Regional Paths of Destruction and Regeneration in the Late 20th Century" Columbia University Press ISBN 0-231-13195-X
- Wald, Matthew L. (17 November 2013). "For First Time, E.P.A. Proposes Reducing Ethanol Requirement for Gas Mix". New York Times. Retrieved 09, March 2020

- Daniel B. Botkin (2001). No man's garden: Thoreau and a new vision for civilization and nature. Island Press. pp. 246–247. ISBN 978-1-55963-465-6. Retrieved 13, June 2020
- "Stolen Goods: The EU's complicity in illegal tropical deforestation" (PDF). Forests and the European Union Resource Network. March 17, 2015. Retrieved 19, January 2020
- "A Policy Framework for Climate and Energy in the Period From 2020 to 2030". European Commission. January 22, 2014. Retrieved 27, May 2020
- "Biofuels: Worried about competition, chemical makers jump into debate". Peterka, Amanda, Greenwire. February 13, 2014. Retrieved 11, June 2020
- EPA, US EPA. "2014 Renewable Fuel Standards for Renewable Fuel Standard program (RFS2): Notice of Proposed Rulemaking". Renewable Fuels: Regulations & Standards. United States Environmental Protection Agency
- "Greenhouse Gas and Energy Life Cycle Assessment of Pine Chemicals Derived from Crude Tall Oil and Their Substitutes". conducted by Franklin Associates, a Division of Eastern Research Group. August 2013. Retrieved 15, July 2020
- John F. Mongillo; Linda Zierdt-Warshaw (2000). Linda Zierdt-Warshaw, ed. Encyclopedia of environmental science. University of Rochester Press. p. 104. ISBN 978-1-57356-147-1

Permissions

All chapters in this book are published with permission under the Creative Commons Attribution Share Alike License or equivalent. Every chapter published in this book has been scrutinized by our experts. Their significance has been extensively debated. The topics covered herein carry significant information for a comprehensive understanding. They may even be implemented as practical applications or may be referred to as a beginning point for further studies.

We would like to thank the editorial team for lending their expertise to make the book truly unique. They have played a crucial role in the development of this book. Without their invaluable contributions this book wouldn't have been possible. They have made vital efforts to compile up to date information on the varied aspects of this subject to make this book a valuable addition to the collection of many professionals and students.

This book was conceptualized with the vision of imparting up-to-date and integrated information in this field. To ensure the same, a matchless editorial board was set up. Every individual on the board went through rigorous rounds of assessment to prove their worth. After which they invested a large part of their time researching and compiling the most relevant data for our readers.

The editorial board has been involved in producing this book since its inception. They have spent rigorous hours researching and exploring the diverse topics which have resulted in the successful publishing of this book. They have passed on their knowledge of decades through this book. To expedite this challenging task, the publisher supported the team at every step. A small team of assistant editors was also appointed to further simplify the editing procedure and attain best results for the readers.

Apart from the editorial board, the designing team has also invested a significant amount of their time in understanding the subject and creating the most relevant covers. They scrutinized every image to scout for the most suitable representation of the subject and create an appropriate cover for the book.

The publishing team has been an ardent support to the editorial, designing and production team. Their endless efforts to recruit the best for this project, has resulted in the accomplishment of this book. They are a veteran in the field of academics and their pool of knowledge is as vast as their experience in printing. Their expertise and guidance has proved useful at every step. Their uncompromising quality standards have made this book an exceptional effort. Their encouragement from time to time has been an inspiration for everyone.

The publisher and the editorial board hope that this book will prove to be a valuable piece of knowledge for students, practitioners and scholars across the globe.

Index

www.ingramcontent.com/pod-product-compliance
Lightning Source LLC
LaVergne TN
LVHW080635210926
848485LV00003B/265